APPLIED AND COMPUTATIONAL STATISTICS
A First Course

THE UNIVERSITY OF WEST LONDON

ELLIS HORWOOD SERIES IN
MATHEMATICS AND ITS APPLICATIONS
Series Editor: Professor G. M. BELL, Chelsea College, University of London

Statistics and Operational Research
Editor: B. W. CONOLLY, Chelsea College, University of London

*In preparation

APPLIED AND COMPUTATIONAL STATISTICS

A First Course

K. D. C. STOODLEY, B.Sc., Ph.D.
School of Mathematical Sciences
University of Bradford

ELLIS HORWOOD LIMITED
Publishers · Chichester

Halsted Press: a division of
JOHN WILEY & SONS
New York · Chichester · Brisbane · Toronto

First published in 1984 by

ELLIS HORWOOD LIMITED
Market Cross House, Cooper Street, Chichester, West Sussex, PO19 1EB, England

The publisher's colophon is reproduced from James Gillison's drawing of the ancient Market Cross, Chichester.

Distributors:

Australia, New Zealand, South-east Asia:
Jacaranda-Wiley Ltd., Jacaranda Press,
JOHN WILEY & SONS INC.,
G.P.O. Box 859, Brisbane, Queensland 40001, Australia

Canada:
JOHN WILEY & SONS CANADA LIMITED
22 Worcester Road, Rexdale, Ontario, Canada.

Europe, Africa:
JOHN WILEY & SONS LIMITED
Baffins Lane, Chichester, West Sussex, England.

North and South America and the rest of the world:
Halsted Press: a division of
JOHN WILEY & SONS
605 Third Avenue, New York, N.Y. 10016, U.S.A.

© 1984 K.D.C. Stoodley/Ellis Horwood Limited

British Library Cataloguing in Publication Data
Stoodley, Keith D.C.
Applied and computational statistics: a first course. –
(Ellis Horwood series in mathematics and its applications)
1. Mathematical statistics
I. Title
519.5   QA276
Library of Congress Card No. 84-12840
ISBN 0-85312-572-4 (Ellis Horwood Limited – Library Edn.)
ISBN 0-85312-805-7 (Ellis Horwood Limited – Student Edn.)
ISBN 0-470-20114-2 (Halsted Press)

Typeset by Ellis Horwood Limited.
Printed in Great Britain by R.J. Acford, Chichester.

Table of Contents

8 **Table of Contents**

Author's Preface

The material in this book constitutes a first course in computational and applied statistics equivalent to approximately twenty-five hours of lectures. Mathematical derivations have been omitted from the main body of the text, although the derivation of some of the more straightforward results is carried out in chapter appendices. Thus, the book is aimed principally at students who study statistics as an ancillary subject in order to apply statistical techniques in their own field. It should, nevertheless, be of use as background reading to those following more specialist mathematical or statistical courses.

The final section of each chapter contains programs developed on desk-top computers to illustrate the material introduced in that chapter. For these sections, it is assumed that the reader is able to program in BASIC. However, although the computing section refers back to the remainder of the chapter, the converse is not true. Thus, the reader who cannot program in BASIC can still make good use of the book. His loss will be his inability to appreciate the impact of the microcomputer in reducing the arithmetical rigours of statistical calculations, and the facilities provided by such machines for simulating random data and experimental situations.

Throughout the text, emphasis is placed on the concepts underlying the techniques discussed, the conditions under which the techniques are valid and the models assumed for the various experimental situations under investigation. The text is liberally illustrated with worked examples. Stress is placed on the use of calculators.

Each chapter is concluded with a set of exercises, to which answers are given. The exercises are divided into three sections; those in Section A are drawn from applications in the life sciences, while those in Section B relate to the physical sciences and engineering applications. The exercises in Section C are computer-orientated.

The book serves as an introductory text to *Applied Statistical Techniques* (Ellis Horwood, 1980) by Stoodley, Lewis and Stainton, and also to other more advanced statistical texts.

In conclusion, I should like to thank Mrs V. M. Hunter for her skilful typing of the manuscript and her equal skill in interpreting my handwriting; Mrs Jennifer Braithwaite for preparing the figures in the text; the University of Bradford for permission to use material from past examination papers; and the Biometrika Trustees for permission to make use of material from *Biometrika Tables for Statisticians,* Volume I, in preparing Tables A1, A2, A3 and A4 of the Appendix.

The computer programs listed in the text are available in substantially the same form on disk. Copies of the disk may be obtained from the publisher, Ellis Horwood Limited, Market Cross Hous, Cooper Street, Chichester, West Sussex, PO19 1EB.

Although the programs in this book have been carefully tested, it is the user's responsibility to ensure that normal checks are made in any application to which the programs are put.

Glossary of Greek characters used in the text

Greek character	English equivalent	Read as	Usage	Chapter
α	a	alpha	provisional mean	2.6
			Poisson distribution	4.4
			significance level	5
			line intercept	6
β	b	beta	line gradient	6
δ	d	delta	small increment	4.5.1
			term in model	5.8.2
ϵ	e	epsilon	experimental error	5.8.2, 6
λ	l	lambda	Poisson mean	4.4
μ	m	mu	population mean	all
			negative exponential distribution	4.5.2
ν	n	nu	degrees of freedom	5, 6
ρ	r	rho	population correlation coefficient	6.3
σ	s	sigma	population standard deviation	all
ξ	x	xi	coded variable	2.6
χ	ch	chi	chi-square distribution	5.5, 5.9
Σ	S	sigma	summation sign	all

To S.K.S.

1

Introduction

1.1 INFERENCE AND DECISION MAKING IN THE FACE OF UNCERTAINTY

Statistics is a highly logical and precise way
of saying half-truths inaccurately.

Statistics is like dynamite — correctly used by experts it can be a useful and constructive tool; in careless or unscrupulous hands it can be a dangerous weapon. Used properly, statistics will allow an experimenter to quantify his concepts and conclusions, and help him to design his experiments so as to take into account sources of systematic variation and to minimise the effect of random error. It will draw his attention to the accuracy of his data and to the type and quality of the inferences which can be made from them. It will enable him to make rational decisions on the basis of a well established range of techniques, in spite of any random component of variation which may be present in his data. In this book the most important of the basic techniques used in modern statistics will be introduced; at the same time the limitations of the methods and the conditions under which they are valid will be stressed. It is hoped that in this way the reader will not be tempted to apply the methods in inappropriate circumstances.

The word 'statistics' is used in several distinct senses which may be summarised as follows:

(a) 'Statistics' may refer to a collection of data or observations, for example, accident statistics or vital statistics.
(b) 'Statistics' may refer to a body of techniques which have been developed in order to analyse data. The general object of such an analysis is to extract the maximum amount of relevant information from the data.
(c) A 'statistic' (where the word is used here in the singular) refers to a function of the observations (as, for example, the mean of the observations (Chapter 2)) which summarises some aspect of the information contained by the data.

Usually the particular interpretation to be placed on the word statistics will be obvious from the context in which it is used.

The first step in the analysis of a large body of data (for example the weights of 100 aspirin tablets illustrated in Table 2.1) is to summarise the information in the data by the calculation of appropriate statistics. In Chapter 2 it will be shown how to calculate statistics to illustrate the central tendency of the data and the dispersion of the data about this central value. Various methods of graphical presentation of the data, which help to elucidate their underlying structure, will also be described.

In addition to calculating statistics which will satisfactorily summarise the data we often require to make inferences, using the information in a sample, about the population from which the sample is drawn.

In quantitative experiments carried out in physics, engineering, chemistry, biology and many other disciplines, the responses obtained are often subject to random, as well as systematic, experimental errors. Such errors can arise in either, or both, of two different ways

(a) because of limitations in the accuracy of the apparatus being used and/or in the abilities of the experimenter,
(b) because of the nature of the law under investigation.

As an illustration of the situation (a) consider an experiment which involves the measurement of a time interval using a stop clock. Then, if the experimenter measured the interval several times as accurately as possible using the same stop clock, his results would vary over a range of values because of variation in his own response, limitations in the accuracy to which the clock can be read and possible sources of inaccuracy in the mechanism of the clock. Some of the errors may be systematic; for example a particular experimenter may consistently be a fraction late in starting the clock or the clock may be running fast; other errors will be random. The first step in the elimination of error in an experiment is to use the best methods and apparatus available and to perform the experiment carefully. Statistics will then allow us to cope in the most efficient manner with sources of error which are left.

As an example of situation (b) consider an experiment to investigate the relationship between the blood pressure and age of a person. If we take a random sample of adult males and plot their blood pressure against their ages it will be evident that there is an approximate linear relationship between these two variables. In this situation the approximate nature of the relationship does not arise to any appreciable extent from experimental errors in the measurement of blood pressure (although these may well exist) or age, but rather from the statistical nature of the underlying 'law' relating blood pressure to age.

The techniques of *statistical inference* enable deductions to be made concerning the parameters of the population from which a sample is drawn, when the observations are subject to random experimental error. Examples of situations in which such techniques may be applied are

(a) The weights of a random sample of 100 aspirin tablets from a large batch are known. We would like to estimate the mean weight of the tablets in the batch and to give some indication of the accuracy of our estimate.

(b) A large batch of bottles has been filled with fluid by an automatic process. On the basis of a random sample of the filled bottles we would like to test whether the mean volume of the fluid in the bottles from the batch differs significantly from the nominal content of the bottles.

(c) Patients suffering from a complaint are split into pairs matched as far as possible according to factors such as age, sex and medical history. One member of each pair is selected at random and is given a control drug, and the other member is given a new drug. We would like to know if the new drug is more effective than the control drug in the treatment of the complaint.

(d) A number of operators measure the percentage of a component in a mixed fluid, each operator using two methods. We would like to know whether there is a bias between the methods.

(e) The blood pressures of a random sample of adult males are measured. We would like to establish a relationship between blood pressure and age.

Various techniques for making statistical inferences are introduced in Chapter 5, where they are applied to situations such as (a) to (d) above. These techniques are based on the distributions associated with the random variables being measured, and the properties of several important distributions are described in Chapter 4. The consideration of these properties in turn requires the concept of probability and laws for the combination of probabilities of events. These topics are discussed in Chapter 3.

In situation (e) above we are faced with the problem of finding the 'best' straight line through a series of experimental points. This topic will be discussed in Chapter 6.

1.2 STATISTICAL COMPUTING

To err is human – it needs a computer to really foul things up.

The modern digital computer has become an invaluable aid to the statistician. Indeed many statistical techniques developed theoretically in the 1930s have only recently become of practical use with the advent of the computer. The desk-top computer, made possible by microprocessors which compress complex circuitry into an exceedingly small space, is more than adequate for carrying out the calculations associated with the techniques discussed in this book. Each chapter is illustrated by a final section which contains programs, developed on a desk-top computer, relevant to the material in the chapter. To make use of these programs, and the corresponding section of the exercises for the chapter (section C), the reader must have a knowledge of the programming language BASIC. Other readers may omit the computing sections, as the remainder of the book does not depend on them; however, they will miss the opportunity of seeing how the computer can ease the numerical work involved in statistical calculations, and also of generating their own data in order to illustrate and test the techniques developed in the book.

1.3 SCALES OF MEASUREMENT

We conclude this introductory chapter with a note on the types of data which may be subjected to statistical analysis. Data may be classified according to their scales of measurement as follows

(a) Nominal scale
On a *nominal scale* numbers are used as labels and no ranking is possible; for example groups of people could be labelled 1, 2 and 3 corresponding to the classification Liberal, Socialist and Conservative.

(b) Ordinal scale
On an *ordinal scale* scores can be assigned to the observations in such a way that the rankings obtained are meaningful, but the intervals between the rankings need not be defined; for example in a survey people's opinion of a product may be ranked according to the scale (1) poor, (2) fairly good, (3) good, (4) very good, (5) excellent.

(c) Interval scale
On an *interval scale* both the ranking and intervals are meaningful, but there is no meaningful zero; an example is the measurement of temperature in degrees Celsius.

(d) Ratio scale
On a *ratio scale* the measurements may be ranked, the intervals are meaningful and there is also a meaningful zero. Thus the ratios of measurements are also meaningful if they are made on a ratio scale. Measurements of quantities such as length, area and time are made on ratio scales, as are counts, such as the number of arrivals of cars at a service station in a given period of time.

In the following chapters we shall be concerned with the statistical analysis of data measured on interval and ratio scales. Statistical inferences may also be made on the basis of nominal or ordinal data, but the necessary techniques are not considered in this text.

2

Descriptive Statistics

2.1 INTRODUCTION

Frequently the first step in the analysis of a sample of observations is to summarise the information contained in the data. This may be done in two ways (a) by calculating *statistics* from the data which characterise certain basic features of the sample, and (b) by graphical techniques. Two important statistics which measure the central tendency and dispersion of the sample are the *mean* and *standard deviation* of the sample respectively. When a large number of observations is available, the reduction of the data to a *frequency distribution* can help to elucidate the structure of the sample. Other useful descriptive statistics are the sample *median, quartiles* and *percentiles.* Graphical techniques include the plotting of *histograms, frequency polygons, bar charts, stem and leaf plots* and *box and dot plots.*

The concepts and methods introduced will to a large extent be illustrated by consideration of the data of Table 2.1. These data are the weights (to the nearest mg) of 100 aspirin tablets as determined using a laboratory balance.

Table 2.1

0.339	0.329	0.337	0.330	0.329	0.330	0.338	0.335	0.331	0.336
0.330	0.333	0.329	0.337	0.334	0.328	0.334	0.333	0.329	0.333
0.332	0.334	0.334	0.330	0.335	0.332	0.335	0.332	0.333	0.335
0.333	0.333	0.337	0.335	0.330	0.328	0.332	0.330	0.331	0.333
0.330	0.330	0.334	0.338	0.327	0.330	0.327	0.331	0.330	0.329
0.329	0.332	0.331	0.330	0.328	0.328	0.328	0.328	0.336	0.334
0.329	0.329	0.332	0.327	0.327	0.333	0.337	0.333	0.325	0.324
0.328	0.331	0.330	0.332	0.335	0.332	0.332	0.325	0.329	0.331
0.334	0.329	0.327	0.334	0.331	0.332	0.332	0.327	0.337	0.332
0.328	0.331	0.327	0.332	0.331	0.330	0.335	0.330	0.332	0.332

2.2 MEAN VALUE

Consider a sample of n observations, denoted by $x_1, x_2, \ldots, x_n$, on a variable X. The variable X could for example be the weight of a tablet, a dimension of a manufactured article, time to failure of a light bulb, yield of a product in a chemical reaction, or the number of radioactive disintegrations in a given period. The *mean* $\bar{x}$ of the sample is a statistic which measures the central tendency of the sample. It is defined by

$$\bar{x} = \frac{1}{n}(x_1 + x_2 + \ldots + x_n)$$

$$= \frac{1}{n}\sum_{i=1}^{n} x_i \qquad\qquad (2.1)$$

The notation $\sum_{i=1}^{n} x_i$ is a shorthand way of writing the sum $x_1 + x_2 + \ldots + x_n$.

The notation tells us to sum the values of x, starting at x_1 ($i = 1$) and finishing at x_n ($i = n$).

The sample mean $\bar{x}$ is useful not only as a descriptive statistic, but is also a basic statistic for many important techniques of statistical inference (Chapter 5).

Sample means will be denoted by $\bar{x}$, while the means of the populations from which the samples are drawn will be denoted by the letter μ (Greek 'mu').

The calculation of the sample mean is illustrated in the examples below.

Ex. 2.1. A sample of ten aspirin tablets is chosen at random and each tablet is weighed. The results (in grams) are

$$0.339, 0.329, 0.337, 0.330, 0.329, 0.330, 0.338, 0.335, 0.331, 0.336$$

Find the mean weight of the sample.

x	
0.329	The mean weight of the sample is
0.329	$\bar{x} = 0.333$ g
0.330	When carrying out the calculations by hand the weights are
0.330	put down in column form, preferably in ascending order; using
0.331	an electronic calculator this is not necessary since the results
0.335	may be entered directly into the calculator.
0.336	In calculators with statistical functions, the addition of the
0.337	observations and division by n is automatically carried out
0.338	within the calculator. The resulting mean value may be immedia-
0.339	tely displayed by pressing the appropriate button on the
10)3.334	keyboard.
0.333 (4)	

Ex. 2.2. An experiment involves determining the increase (+) or decrease (−) in weights of a sample of laboratory animals. In a particular experiment involving 6 animals the changes in weight (to the nearest gram) are

$$123, -15, 65, 24, -76, -3$$

Calculate the mean change in weight.

x	
−76	
−15	
−3	
	24
	65
	123
−94	212
	−94
6	118
	19.6 $\bar{x}$

In carrying out the calculations by hand it is easier to sum the positive and negative values separately and to take the algebraic sum of the totals.

In using an electronic calculator it is just as easy to enter the values in the order in which they are obtained, irrespective of sign.

2.3 FREQUENCY DISTRIBUTIONS

Here we consider the situation in which we have n observations but these involve only $k(<n)$ distinct values, denoted by $x_1, \ldots, x_k$, of the observed variable X. This is illustrated in the example below.

Ex. 2.3. A sample of 20 aspirin tablets is taken and their weights determined to the nearest 0.005 g giving

0.340, 0.330, 0.335, 0.330, 0.330, 0.330, 0.340, 0.335, 0.330, 0.335
0.330, 0.335, 0.330, 0.335, 0.335, 0.330, 0.335, 0.335, 0.330, 0.335

Find the sample mean.

We could of course proceed to find the sample mean using the methods introduced in the previous section. However, we notice that although we have $n = 20$ measurements there are in fact only $k = 3$ distinct values of X, $x_1 = 0.330$, $x_2 = 0.335$ and $x_3 = 0.340$. We reduce the sample to a *frequency distribution* by counting the number f_i of observations which have the value x_i to give

x_i	0.330	0.335	0.340
f_i	9	9	2

Then $\bar{x} = \dfrac{1}{20} \Big((0.330 + 0.330 + \ldots + 0.330) + (0.335 + \ldots + 0.335)$

$$+ (0.340 + 0.340) \Big)$$

$$= \frac{1}{20}(9 \times 0.330 + 9 \times 0.335 + 2 \times 0.340)$$

$$= 0.333(25)$$

In general if we have a set of values $x_1, x_2, \ldots, x_k$, where x_i occurs with frequency $f_i(i = 1, \ldots, k)$ and $f_1 + f_2 + \ldots + f_k = n$, then

$$\bar{x} = \frac{1}{n} \sum_{i=1}^{k} f_i x_i \qquad (2.2)$$

Note that the set of weights in example 2.3 form a frequency distribution only because they have been rounded off to the nearest 0.005 g. Since weight is a continuous variable, by increasing the accuracy with which the weighings are made a stage will eventually be reached in which each of the aspirin tablets will have a different weight (to the number of significant figures concerned) and the sample will no longer form a frequency distribution.

A frequency distribution which is formed from classes covering ranges of a continuous variable, or more than one value of a discrete variable, is referred to as a *grouped* frequency distribution. In order to elucidate the structure of a large sample we can reduce the range covered by the sample to a number of classes, usually of equal length, and then count the number of members of the sample falling into each class; in this way the sample is reduced to a grouped frequency distribution.

If the variable under observation is discrete the sample will fall naturally into a frequency distribution according to the set of values which the variable can take. For example suppose that we are counting the number of cars per household, X, in a survey. Suppose that 5 is the maximum value of X encountered during the survey. Then the sample surveyed will automatically reduce to a frequency distribution with six classes ($X = 0, 1, 2, 3, 4, 5$) and it would not be possible to further subdivide these classes based only on consideration of the numbers of cars per household. However, it would be possible to form a grouped distribution; for example we could combine together households with 0 or 1 car, 2 or 3 cars, and 4 or 5 cars, to form a gouped frequency distribution with three classes.

In an *ungrouped* frequency distribution all the members in a given class have exactly the value of the observed variable corresponding to that class. In a grouped frequency distribution each member in a given class is assigned a value corresponding to the mid-point of the range covered by the class. In the case of the data for example 2.3 the class with mid-value $x_2 = 0.335$ will actually cover a range of $0.3325 < X < 0.3375$ since the results have been rounded to the nearest 0.005 g.

The number of classes taken in forming a grouped frequency distribution is to some extent arbitrary. As the class width is reduced a situation will eventually be reached in which there is a frequency of either 1 or 0 in any class. On the other hand if the class width is increased the range of the observed variable covered by the observations in the sample will fall into fewer and fewer classes and the picture of the structure of the sample presented will become cruder and less informative. As a compromise between these extremes we would, for example, with a sample of size 100, aim at forming a frequency distribution with about 10 classes.

Usually the class widths are taken to be all the same but this is not essential. If the tails of the distribution contain only a few members of the sample it may be convenient to take wider classes in the tails of the distribution than in the remainder of the range.

For a grouped frequency distribution the value obtained for the sample mean using (2.2) will depend on the width of the classes used. However, this variation will in general be small and for practical purposes can be ignored.

Ex. 2.4. Reduce the weights of 100 tablets illustrated in Table 2.1 to a frequency distribution and calculate the mean weight of the tablets.

We first look for the lowest and highest values in the sample. These are 0.324 g and 0.339 g. We split this range up into 8 classes, the first including members of the sample with weights 0.324 and 0.325 g, the second 0.326 and 0.327 g, . . . , the eighth 0.338 and 0.339 g. Note that the weight ranges covered by the classes are 0.3235–0.3255 g, 0.3255–0.3275 g and so on, since the weights are all rounded to the nearest milligram. The class mid-values are 0.3245 g, 0.3265 g, ... (If we had taken the class ranges to be 0.323–0.325 g, 0.325–0.327 g, . . . then we would not know whether a tablet with a weight of 0.325 g should be allocated to the first or second classes, whereas with the class boundaries as chosen above there is no such ambiguity.)

We now go through the data systematically assigning the observations to classes. When an observation is assigned to a class a tally mark is made in the frequency distribution table. For ease of addition the tally marks are grouped in fives. As a check the class frequencies should sum to the total frequency ($\equiv$ sample size). However, this is only a partial check as it does not allow for allocation of a sample member to the wrong class.

Weights in class	Class boundaries	Class mid-value (x)	Tally marks	Frequency (f)
0.324–0.325	0.3235–0.3255	0.3245	111	3
0.326–0.327	0.3255–0.3275	0.3265	JHT 11	7
0.328–0.329	0.3275–0.3295	0.3285	JHT JHT JHT 111	18
0.330–0.331	0.3295–0.3315	0.3305	JHT JHT JHT JHT 111	23
0.332–0.333	0.3315–0.3335	0.3325	JHT JHT JHT JHT 1111	24
0.334–0.335	0.3335–0.3355	0.3345	JHT JHT JHT	15
0.336–0.337	0.3355–0.3375	0.3365	JHT 11	7
0.338–0.339	0.3375–0.3395	0.3385	111	3
				100

The mean value of the weights of the tablets in the sample is calculated by attributing to each member in a given class, the class mid-value x_i.

Then
$$\bar{x} = \frac{1}{n} \sum_{i=1}^{k} f_i x_i$$

$$= \frac{1}{100} \sum_{i=1}^{8} f_i x_i$$

$$= \frac{1}{100} (3 \times 0.3245 + \ldots + 3 \times 0.3385)$$

$$= \frac{1}{100} \times 33.142$$

$$= 0.33142 \ .$$

2.4 GRAPHICAL REPRESENTATION OF A FREQUENCY DISTRIBUTION

In this section the following methods of graphical representation of frequency distributions will be considered (a) stem and leaf plot (b) histogram (c) frequency polygon (d) bar chart. Later (section 2.8.2) the construction of the box and dot plot will be described.

2.4.1 Stem and leaf plot

The numerical values of the observations in each class of the distribution are divided into two parts (a) the *stem* which consists of all the digits common to the members of the class, and (b) the *leaf* which consists of the remaining digits. The stems are then ordered with the smallest at the top and the leaves are written on the same lines as their stems to give a working *stem and leaf plot*. Finally the leaves in each of the rows may be put into ascending numerical order to give an ordered stem and leaf plot.

Ex. 2.5. Construct working and ordered stem and leaf plots for the data of example 2.4.

(i) Working stem and leaf plot

Class		
Stem	Leaves	
0.32	(4,5)	545
0.32	(6,7)	77777 77
0.32	(8,9)	99989 89988 88998 998
0.33	(0,1)	00100 00100 01010 10111 100
0.33	(2,3)	33322 23332 32233 22222 2222
0.33	(4,5)	54444 55554 45445
0.33	(6,7)	76776 77
0.33	(8,9)	988

(ii) Ordered stem and leaf plot

Class Stem	Leaves	
0.32	(4,5)	455
0.32	(6,7)	77777 77
0.32	(8,9)	88888 88899 99999 999
0.33	(0,1)	00000 00000 00001 11111 111
0.33	(2,3)	22222 22222 22222 33333 3333
0.33	(4,5)	44444 44455 55555
0.33	(6,7)	66777 77
0.33	(8,9)	889

It has been assumed in the above description that the leaves are the set, or a subset of the digits 0, 1, . . . , 9. Program 2.3 (section 2.9) allows the leaves to range from 0 to 99.

2.4.2 Histogram

The *histogram* is used for the graphical representation of grouped frequency distributions. Figure 2.1 illustrates the construction of the histogram for the data of example 2.4. The vertical axis represents the class frequency, and the

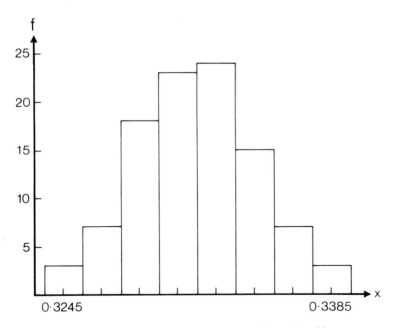

Fig. 2.1 – Histogram for the weights of 100 aspirin tablets.

class mid-values x_i ($i = 1, 2, \ldots, 8$) are plotted along the horizontal axis. With the class mid-value x_i as the centre of its base, a rectangle, of base equal to the class width and height equal to f_i, is erected for each of the classes.

If the class widths δx_i are not all equal, a histogram constructed by the above method will give a distorted picture of the distribution. It will over-emphasise the contributions of the classes with the larger widths. In this situation the histogram is constructed with $f_i/\delta x_i$ along the vertical axis instead of f_i. When the δx_i are all the same ($= \delta x$, say) the shape of the histogram will be the same whether f_i or $f_i/\delta x$ is plotted against x_i.

2.4.3 Frequency polygon

An alternative method for the graphical representation of a grouped frequency distribution is the *frequency polygon*. This is illustrated in Fig. 2.2 for the data of example 2.4. In the case of equal class widths the points (x_i, f_i) are plotted. These points are then joined by straight lines to form an open polygon which is referred to as the frequency polygon. If the class widths are not all equal the construction is based on the points $(x_i, f_i/\delta x_i)$ as in the case of the histogram.

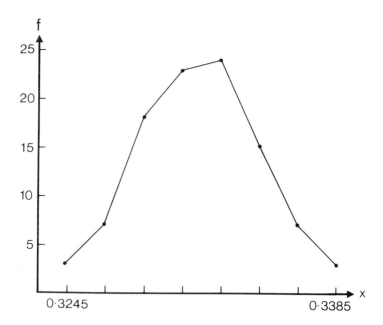

Fig. 2.2 – Frequency polygon for the weights of 100 aspirin tablets.

2.4.4 Bar chart

A *bar chart* is used for the graphical representation of frequency distributions in which all the members in a given class have the same value. Here the class values are plotted along the x-axis and a vertical line (or bar) of height equal to the class frequency is erected over the class value.

Ex. 2.6. The number of defectives in one hundred samples each of 60 plastic beakers are noted during a period in which the manufacturing process is under control in order to set up a control chart. The distribution of defectives amongst the samples is tabulated below.

X	0	1	2	3	4	5	6
f	11	32	26	14	12	4	1

Illustrate the distribution by means of a bar chart.

The chart is shown in Fig. 2.3.

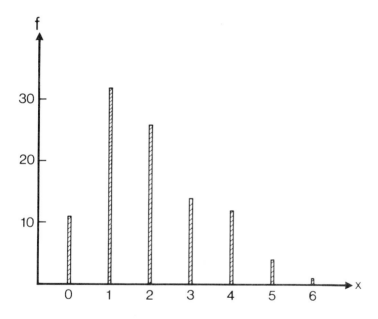

Fig. 2.3 – Bar chart for the distribution of defectives in one hundred samples.

2.5 STANDARD DEVIATION AND VARIANCE

The mean alone is not sufficient to summarise all the important information in a sample of observations. A second statistic is required to give a measure of the spread of the observations about their mean value. For instance consider two samples, each of three tablets; let the weights of the tablets in the first sample be 0.325, 0.326 and 0.330 g and in the second be 0.320, 0.326 and 0.335 g. Then the mean of each sample is 0.327 g. However, the spread of the sample values about the mean is clearly different for the two samples. For the first sample the deviations of the sample values from the mean are $-0.002, -0.001$ and 0.003 g, and for the second the deviations are $-0.007, -0.001$ and 0.008 g.

We notice that in each case the algebraic sum of the deviations from the mean is zero; this is in fact a general result which follows from the definition of the mean. Hence the sum of the deviations cannot be directly used to measure the dispersion of the sample values. Instead each deviation $x_i - \bar{x}$ is first squared so that the resulting quantities $(x_i - \bar{x})^2$ are all positive. The *sample variance* s^2 is then defined by

$$s^2 = \frac{1}{n-1} \sum_{i=1}^{n} (x_i - \bar{x})^2 \qquad (2.3)$$

The divisor $(n-1)$ is used instead of n here as it can be shown that the sample variance s^2 can then be used without adjustment as an estimate of the population variance σ^2 (Greek 'sigma'). This result proves very useful in the context of statistical inference (Chapter 5).

Where the data $x_1, \ldots, x_n$ represent the whole population under consideration, rather than a sample from the population, the population variance σ^2 is defined by

$$\sigma^2 = \frac{1}{n} \sum_{i=1}^{n} (x_i - \mu)^2 \qquad (2.4)$$

where the population mean is $\mu = \dfrac{1}{n} \sum_{i=1}^{n} x_i$.

An alternative measure of dispersion is the *standard deviation* of a sample, where the standard deviation s is the square root of the variance s^2. The standard deviation has the advantage of being measured in the same units as the data from which it is derived. In practice each of the two measures is frequently encountered.

Calculators with statistical functions will display the standard deviation of a sample on entering the sample observations into the calculator. The calculator manual should be consulted to ascertain whether the variance is being calculated using a divisor of $n-1$ (equation (2.3)) or n (equation (2.4)). Some calculators provide both options via keys labelled 'σ_{n-1}' and 'σ_n' respectively.

For a frequency distribution with class values $x_1, x_2, \ldots, x_k$ and class frequencies $f_1, \ldots, f_k$ (with $f_1 + \ldots + f_k = n$) the variance s^2 is defined by

$$s^2 = \frac{1}{n-1} \sum_{i=1}^{k} f_i (x_i - \bar{x})^2 \qquad (2.5)$$

Note that (2.5) may be obtained from (2.3) by grouping together the f_1 observations with value x_1, the f_2 observations with value x_2 and so on.

In addition to its value as a descriptive statistic, the sample variance, like the sample mean, also plays an important role in problems of statistical inference.

Some algebra (see Appendix 2.1) shows that the formulae (2.3) and (2.5) may be written respectively in the alternative forms

$$s^2 = \frac{1}{n-1} \left\{ \sum_{i=1}^{n} x_i^2 - \frac{1}{n} \left(\sum_{i=1}^{n} x_i \right)^2 \right\} \qquad (2.6)$$

and

$$s^2 = \frac{1}{n-1} \left\{ \sum_{i=1}^{k} f_i x_i^2 - \frac{1}{n} \left(\sum_{i=1}^{k} f_i x_i \right)^2 \right\} \qquad (2.7)$$

In general formulae (2.6) and (2.7) are more convenient from a computational point of view. However when the observations are closely grouped the terms

$$\left(\sum_{i=1}^{n} x_i^2 \text{ and } \frac{1}{n} \left(\sum_{i=1}^{n} x_i \right)^2 \text{ in (2.6)} \quad \left(\text{or } \sum_{i=1}^{k} f_i x_i^2 \text{ and } \frac{1}{n} \left(\sum_{i=1}^{k} f_i x_i \right)^2 \text{ in (2.7)} \right) \right.$$

can be very nearly equal; this can lead to the loss of an unacceptably large number of significant figures when the terms are subtracted. This can be avoided by the use of coding (section 2.6) or by first calculating $\bar{x}$ and then using formula (2.3) or (2.5).

Ex. 2.7. Find the mean, variance and standard deviation of the observations

$$0.416 \qquad 0.442 \qquad 0.444 \qquad 0.446 \qquad 0.469.$$

We use the formula

$$s^2 = \frac{1}{n-1} \left\{ \sum_{i=1}^{n} x_i^2 - \frac{\left(\sum_{i=1}^{n} x_i \right)^2}{n} \right\} .$$

x	x^2
0.416	0.173056
0.442	0.195364
0.444	0.197136
0.446	0.198916
0.469	0.219961

$5\overline{|2.217}\ \Sigma x \qquad 0.984433\ \Sigma x^2$

$0.4434\ \bar{x} \qquad 0.9830178\ (\Sigma x)^2/n$

$4\overline{|0.0014152}$

$0.0003538\ s^2$

$0.018810\ \ s$

Because the sample values are closely grouped, $\sum\limits_{i=1}^{n} x_i^2$ and $\dfrac{1}{n}\left(\sum\limits_{i=1}^{n} x_i\right)^2$ are nearly equal. Hence no rounding off should be carried out during the intermediate stages of the calculation.

    In general there should be no difficulty in obtaining the standard deviation with sufficient accuracy from scientific calculators, even when the observations are closely grouped.

Ex. 2.8. Water is run from a burette up to the 10 ml mark in a 10 ml measure by each of the members of a class and the differences between the initial and final burette readings obtained. The results (in ml) obtained by the class are tabulated below:

9.6	10.0	9.9	9.6	10.0	9.5	10.0	9.7	10.2	10.0
9.8	9.8	9.9	9.8	9.8	9.9	9.6	9.8	10.3	9.6
9.7	9.8	10.1	10.1	10.1	9.9	9.9	9.8	9.5	10.0
9.8	10.3	9.7	9.8	9.8	10.1	9.9	9.4	9.9	9.9
10.0	10.1	10.2	9.8	10.1	10.0	10.1	9.9	9.9	10.1
10.0	10.2	9.9	10.0	10.0					

Reduce the sample to a frequency distribution and calculate the mean and variance of the distribution.

Choosing class mid-values $x = 9.4, 9.5, \ldots$ the sample may be reduced to a distribution as follows:

x	Tally marks	f	fx	fx^2
9.4	1	1	9.4	88.36
9.5	11	2	19.0	180.50
9.6	1111	4	38.4	368.64
9.7	111	3	29.1	282.27
9.8	L⊦⊦1 L⊦⊦1 1	11	107.8	1056.44
9.9	L⊦⊦1 L⊦⊦1 1	11	108.9	1078.11
10.0	L⊦⊦1 L⊦⊦1	10	100.0	1000.00
10.1	L⊦⊦1 111	8	80.8	816.08
10.2	111	3	30.6	312.12
10.3	11	2	20.6	212.18

 55 n 55 | 544.6 Σfx 5394.70 Σfx^2

                          5392.530182 $(\Sigma fx)^2/$

             9.90182

                54 | 2.169818

                     0.040182 s^2

                     0.2005 s

Hence the sample mean is 9.90 ml with a standard deviation of 0.20 ml.

2.6 CODING

When the observations in a sample are closely grouped, the structure of the sample can be clarified by the use of *coding*. The first step in coding is to subtract a *provisional mean* α from each of the observations (or class mid-values in the case of a frequency distribution). α could for example be chosen as the lowest value in the sample, or as a convenient value near the estimated sample mean.

We can then write

$$x_i = \xi_i + \alpha \qquad i = 1, 2, \ldots, n \ .$$

Suppose that $\xi_1, \ldots, \xi_n$ have a common factor c; taking out this factor gives

$$\xi_i = cu_i$$

so that

$$x_i = cu_i + \alpha \ . \tag{2.8}$$

From (2.8) it can be shown that (see Appendix 2.2)

$$\bar{x} = c\,\bar{u} + \alpha \tag{2.9}$$

and

$$s_x^2 = c^2\, s_u^2 \ . \tag{2.10}$$

For the distribution of example 2.4 α could be chosen as 0.3305 and c as 0.002 to give values of u of -3, -2, -1, 0, 1, 2, 3, and 4 for the class mid-values; $\alpha = 10.0$ and $c = 0.1$ for the data of example 2.8 leads to values for u of -6, $-5, \ldots, 2, 3$ corresponding to the class values $9.4, \ldots, 10.3$.

Appropriate use of coding can remove the numerical difficulty of having to subtract two nearly equal numbers in obtaining the standard deviation of a closely grouped sample. However, an extra stage is introduced into the calculations.

Ex. 2.9. Repeat the calculations of example 2.7 using coding

x	α	ξ	c	u	u^2
0.416	0.444	−0.028	0.001	−28	784
0.442		−0.002		−2	4
0.444		0.000		0	0
0.446		0.002		2	4
0.469		0.025		25	625

$$5\;\boxed{-3} \qquad 1417$$
$$1.8$$
$$-0.6$$
$$-0.0006 \; c\bar{u} \qquad 4\,\boxed{1415.2}$$
$$0.4440 \; \alpha$$
$$\overline{} \qquad\qquad 353.8 \qquad s_u^2$$
$$0.4434 \; \bar{x} \qquad\qquad 10^{-6} \times c^2$$
$$\overline{}$$
$$353.8 \times 10^{-6} \quad s_x^2$$
$$0.01881 \qquad s_x$$

2.7 PERCENTILES

Sample *percentiles* are descriptive statistics which are sometimes used to supplement the information obtained from the mean and standard deviation. The sample values $x_1, \ldots, x_n$ are first of all arranged in order of ascending magnitude $(x_1 \leqslant x_2 \leqslant \ldots \leqslant x_n)$. Then the pth percentile is defined to be the value of x below which $p\%$ of the sample values lie. The method of obtaining this value is described below (a) for samples not reduced to frequency distributions (b) ungrouped frequency distributions and (c) grouped frequency distributions.

2.7.1 Sample percentiles

As above let the sample values be arranged in order of ascending magnitude $x_1 \leqslant x_2 \leqslant \ldots \leqslant x_n$. Then the pth percentile is x_q where $q = \dfrac{p}{100}(n+1)$. If q is not an integer, but lies between two integers m and $m+1$, the pth percentile may be found by interpolating between x_m and x_{m+1} according to the formula

$$x_q = x_m + (q - m)(x_{m+1} - x_m)$$

The 25th, 50th and 75th percentiles are alternatively called the *first* (or *lower*) *quartile*, *median* (or *second quartile*) and *third* (or *upper*) *quartile* of the sample.

Ex. 2.10. The observations, arranged in ascending order, in a random sample of size 11 are 328, 328, 329, 331, 332, 333, 334, 335, 335, 336, 336. Find the median, first and third quartiles and 60th percentile of the sample.

The median is the value corresponding to the $\dfrac{11 + 1}{2}$ or 6th observation, that is 333.

The first quartile is the value corresponding to the $\dfrac{11 + 1}{4}$ or 3rd observation, that is 329.

The third quartile is the value corresponding to the $\dfrac{3(11 + 1)}{4}$ or 9th observation, that is 335.

The 60th percentile is the value corresponding to the $\dfrac{60(11 + 1)}{100}$ or 7.2nd observation that is $334 + 0.2\,(335 - 334)$

$$= 334.2.$$

2.7.2 Percentiles of an ungrouped frequency distribution

Ex. 2.11. A sample of 20 tablets is a mixture of six different types of tablets containing percentages $x = 1, 2, 3, 4, 5, 6$ of a certain drug X. The distribution of tablets in the sample is shown below

x	1	2	3	4	5	6
f	2	2	5	6	2	3

Find the median and first quartile of the distribution.

We first form the *cumulative frequency* (*cf*) which is the number of observations in the sample up to and including the end of the class concerned.

x	1	2	3	4	5	6
f	2	2	5	6	2	3
cf	2	4	9	15	17	20

The median is the value corresponding to the $\dfrac{20 + 1}{2} = 10.5$th member of the sample when these are arranged in ascending order of x value. As there are 9 observations up to the end of the third class and 15 up to the end of the fourth class the '10.5th' observation will fall into the fourth class and therefore has value $x = 4$.

The first quartile is the value corresponding to the $\dfrac{20 + 1}{4} = 5.25$th member of the sample which from inspection falls into the third class and therefore has value $x = 3$.

2.7.3 Percentiles of a grouped frequency distribution

Consider a grouped frequency distribution consisting of n observations on a continuous variable x. In this case the pth percentile of the distribution is defined to be the value of x corresponding to the $\dfrac{p}{100} n$ th member of the distribution when they are arranged in ascending order. The difference in definition here arises from the manner in which the class boundaries are chosen for a grouped frequency distribution.

Ex. 2.12. Find the median, first and third quartiles, and 95th percentile of the distribution in example 2.4.

The distribution is

x	0.3245	0.3265	0.3285	0.3305	0.3325	0.3345	0.3365	0.3385
f	3	7	18	23	24	15	7	3
cf	3	10	28	51	75	90	97	100

The median is the weight corresponding to the $\dfrac{50}{100} \times 100 = $ 50th member of the sample when they are arranged in ascending order. From the cumulative frequency we see that this is the 22nd member of the class with mid-value 0.3305 and boundaries 0.3295 and 0.3315. Since x is a continuous variable we make the assumption that the 23 members of this class are uniformly spread throughout the range covered by the class

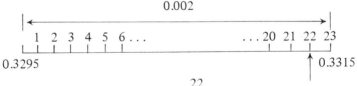

Hence the median value $= 0.3295 + 0.002 \times \dfrac{22}{23} = 0.3295 + 0.0019 = 0.3314$ g.

The first quartile is the weight corresponding to the $\dfrac{1}{4} \times 100 = $ 25th member of the sample which is the 15th member of the third class. Thus

the first quartile $= 0.3275 + \dfrac{15}{18} \times 0.002$

$= 0.3275 + 0.0017$

$= 0.3292,$

$$\text{the third quartile} = 0.3315 + \frac{75 - 51}{24} \times 0.002$$

$$= 0.3315 + 0.0020$$

$$= 0.3335$$

$$\text{and the 95th percentile} = 0.3355 + \frac{95 - 90}{7} \times 0.002$$

$$= 0.3355 + 0.0014$$

$$= 0.3369.$$

The median is an alternative measure of the central tendency of a sample. The value of the median is not so sensitive to outlying observations as the value of the mean. For nearly symmetrical distributions the mean and median are nearly equal in value.

An alternative measure of the dispersion of a sample which is occasionally used is given by the semi-interquartile range (SIQR); this is defined by

$$\text{SIQR} = \frac{\text{third quartile} - \text{first quartile}}{2}$$

2.8 GRAPHICAL TECHNIQUES BASED ON PERCENTILES

2.8.1 Ogives
The percentiles of a grouped frequency distribution may be found graphically from an *ogive* (or *cumulative frequency polygon*). This figure is constructed by plotting the cumulative frequency as y-coordinate against the class boundaries as x-coordinate; these points are then joined by a series of straight lines to form an open polygon. Thus for the distribution of example 2.4 the points plotted would be (0.3235, 0), (0.3255, 3), . . . , (0.3395, 100). The resulting ogive is shown in Fig. 2.4. Then the pth percentile can be read from the ogive as the x-coordinate corresponding to the point on the ogive with y-coordinate $\frac{p}{100} \cdot n$, which is equal to p for this example ($n = 100$). This will lead to the same values for the percentiles as the method of linear interpolation introduced in example 2.12.

2.8.2 Box and dot plots
The *box and dot plot* is a graphical technique for the display of a sample structure based on the median, quartiles and extreme values of the sample. It has the advantage of illustrating the *skewness* of the data and drawing attention to extreme observations (*outliers*). The construction of one form of this plot will be illustrated for the data of Table 2.1. In example 2.12 it was shown that the quartiles of this sample were 0.3292 g, 0.3314 g and 0.3335 g. These values

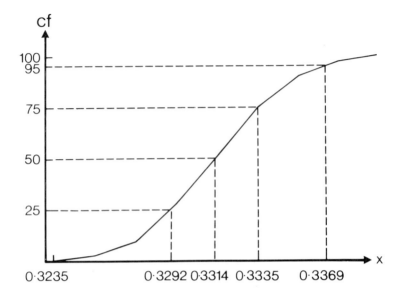

Fig. 2.4 – Ogive for the weights of 100 aspirin tablets.

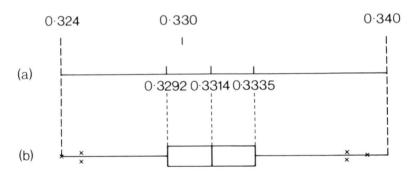

Fig. 2.5 – Box and dot plot for the weights of 100 aspirin tablets.

are first marked on a straight line (Fig. 2.5(a)). Then a narrow rectangular box is drawn (Fig. 2.5(b)) with its narrow ends at the lower and upper quartiles. The position of the median is marked by a vertical line inside the box. The plot is then completed by marking on the line the position of the two extreme values at either end of the ordered sample.

The plot is useful in illustrating the skewness of a sample. If the distribution has a long tail to the right (*positively skew*), the right-hand section of the box will be longer than the left, and the upper extreme points will be further from the median than the lower extreme points. The converse will be true if the distribution is *negatively skew* with its longer tail to the left.

2.9 BASIC PROGRAMS FOR THE EVALUATION OF SAMPLE STATISTICS

Desk-top computers are now an invaluable aid to the user of statistics. The use of microcomputers will be illustrated by the following programs:

(a) Program 2.1   Calculation of the mean, variance and standard deviation of a sample.
(b) Program 2.2   Reduction of data to a grouped frequency distribution; calculation of the mean, standard deviation and percentiles of the distribution.
(c) Program 2.3   Construction of a stem-leaf plot.

It is assumed that the reader is familar with the computing language BASIC. The programs were developed on a Commodore Pet 4016 and may be run directly on a Commodore 64. However, with relatively minor adaptations they can be made compatible with other systems using BASIC. No attempt has been made to include programs needing computer graphics (e.g. plotting histograms) as such programs are usually fairly specific to the system for which they were developed.

The output from the programs is illustrated with the use of data from problems discussed earlier in the chapter. The data are read in from statements incorporated in the programs, but the programs may be readily adapted so that data may be input directly from the keyboard.

One or two general points regarding the programs, which are also relevant to later chapters, should be mentioned at this stage.

(1) The notation used in the programs is either defined in REM statements, or in print-out statements, or is evident from the text.
(2) The practice has been adopted of initially zeroing sums, and arrays which are to be filled. This is strictly speaking, not necessary in most cases as these variables are automatically set to zero by the computer at the beginning of the program.
(3) If the screen is filled by output at an intermediate stage, the following device is used

```
500   PRINT "PRESS ANY KEY TO CONTINUE"
510   GET A$: IF A$ = " " THEN 510
520   PRINT "█": REM ** CLEAR SCREEN
```

The message PRESS ANY KEY TO CONTINUE is printed on the screen. The instruction GET A$ at 510 scans the keyboard for any input. If there is no input (A$ = " ") the program loops at 510, the current visual display being maintained. When a key is pressed instruction 520 is executed, clearing the screen ready for further output.

(4) The following device is used to obtain output when the current value of an index is a multiple of an integer K

     500   IF I = INT(I/K) * K THEN PRINT . . .

For example if K = 20, I = 50 then I/K = 50/20 = 2.5 and INT (I/K) = 2, INT (I/K) * K = 2 $\times$ 20 = 40 $\neq$ I and nothing is printed. If K = 20, I = 60 then I/K = 3 and INT (I/K) = 3, INT (I/K) * K = 3 $\times$ 20 = 60 = I and the print statement is executed.

(5) Numbers are rounded off for output as in the following example

     500   EP = INT(100 * E + 0.5)/100

Suppose E = 10.2334; then 100 * E + 0.5 = 1023.84, INT(100 * E + 0.5) = 1023, and EP = 10.23 giving E rounded to two decimal places. Using the algorithm with different powers of ten will give figures rounded to the appropriate number of decimal places.

Program listings and sample outputs.
(a) Program 2.1

```
100 PRINT"⌂":REM**CLEAR SCREEN
110 PRINT"         **PROGRAM 2.1**
120 PRINT"      CALCULATION OF THE MEAN"
130 PRINT"   VARIANCE AND STANDARD DEVIATION"
140 PRINT"      OF A SAMPLE OF SIZE N"
150 DIM X(100)
160 PRINT:PRINT
170 REM**ZERO SUMS FOR MEAN AND VARIANCE
180 S1X=0 : CXX=0
190 REM**CALCULATION OF MEAN M
200 READ N
210 PRINT"SAMPLE VALUES ARE:-"
220 FOR I=1 TO N
230 READ X(I)
240 PRINT X(I); : IF I=INT(I/10)*10 THEN PRINT
250 S1X=S1X+X(I)
260 NEXT I
270 PRINT:PRINT
280 M=S1X/N
290 REM**CALCULATION OF VARIANCE V AND S.D. S
300 FOR I=1 TO N
310 D=X(I)-M
320 CXX=CXX+D*D
330 NEXT I
340 V=CXX/(N-1) : S=SQR(V)
350 PRINT"SAMPLE MEAN IS ";M
360 PRINT"SAMPLE VARIANCE IS ";V
370 PRINT"SAMPLE STANDARD DEVIATION IS ";S
380 REM**DATA FOR EXAMPLE 2.7
390 DATA 5
400 DATA 0.416,0.442,0.444,0.446,0.469
410 END
```

Output for data of example 2.7

```
SAMPLE VALUES ARE:-
.416  .442  .444  .446  .469

SAMPLE MEAN IS                   .4434
SAMPLE VARIANCE IS               3.53800002E-04
SAMPLE STANDARD DEVIATION IS     .0188095721
```

(b) Program 2.2

```
100 PRINT"◻":REM**CLEAR SCREEN
110 PRINT"        **PROGRAM 2.2**"
120 PRINT"  REDUCTION OF DATA TO A GROUPED"
130 PRINT"    FREQUENCY DISTRIBUTION AND"
140 PRINT"      CALCULATION OF STATISTICS"
150 PRINT:PRINT
160 DIM F(20),CF(20)
170 REM**READ SAMPLE SIZE N,NUMBER OF CLASSES NC,"
180 REM**LOWER BOUND OF FIRST CLASS B,"
190 REM**AND CLASS WIDTH W
200 READ N,NC,B,W
210 REM**READ X VALUES AND REDUCE TO A
220 REM**FREQUENCY DISTRIBUTION
230 REM**F(I) IS THE FREQUENCY IN CLASS I
240 REM**ZERO F'S
250 FOR I=1 TO NC : F(I)=0 : NEXT I
260 PRINT"SAMPLE VALUES ARE:-"
270 FOR I=1 TO N
280 READ XI
290 PRINT XI;
300 IF I=INT(I/10)*10 THEN PRINT
310 REM**FIND CLASS J OF XI
320 J=INT((XI-B)/W)+1
330 F(J)=F(J)+1
340 NEXT I
350 REM**CALCULATION OF MEAN M
360 REM**AND CUMULATIVE FREQUENCIES CF(I)
370 X=B+W/2 : S1X=0 : CF(0)=0
380 FOR I=1 TO NC
390 S1X=S1X+F(I)*X
400 X=X+W
410 CF(I)=CF(I-1)+F(I)
420 NEXT I
430 M=S1X/N
440 REM**CALCULATION OF VARIANCE V
450 X=B+W/2 : CXX=0
460 FOR I=1 TO NC
470 D=X-M
480 CXX=CXX+F(I)*D*D
490 X=X+W
500 NEXT I
510 V=CXX/(N-1)
520 S=SQR(V)
530 PRINT
540 PRINT"PRESS ANY KEY TO CONTINUE"
550 GET A$ : IF A$="" THEN 550
560 PRINT"◻":REM**CLEAR SCREEN
570 PRINT"DISTRIBUTION MEAN IS ";M
```

```
580 PRINT"DISTRIBUTION VARIANCE IS ";V
590 PRINT"DISTRIBUTION S.D. IS ";S
600 PRINT
610 PRINT TAB(10);"CLASS      FREQUENCY"
620 PRINT TAB(10);"MID-VALUE"
630 X=B+W/2
640 FOR I=1 TO NC
650 PRINT TAB(10);X;SPC(6);F(I)
660 X=X+W
670 NEXT I
680 PRINT
690 PRINT"PRESS ANY KEY TO CONTINUE"
700 GET A$ : IF A$="" THEN 700
710 PRINT"◻":REM**CLEAR SCREEN
720 REM**CALCULATION OF PERCENTILES
730 REM**READ IN NUMBER TO BE FOUND
740 READ NP
750 REM**READ IN PERCENTILE P
760 FOR I=1 TO NP
770 READ P
780 Q=N*P/100
790 J=0
800 J=J+1
810 IF Q<CF(J) THEN 830
820 GOTO 800
830 J=J-1
840 XQ=B+J*W+W*(Q-CF(J))/F(J+1)
850 PRINT P;"TH PERCENTILE IS ";XQ
860 NEXT I
870 REM**DATA FROM TABLE 2.1
880 DATA 100,8,0.3235,0.002
890 DATA .339,.329,.337,.330,.329,.330,.338,.335,.331,.336
900 DATA .330,.333,.329,.337,.334,.328,.334,.333,.329,.333
910 DATA .332,.334,.334,.330,.335,.332,.335,.332,.333,.335
920 DATA .333,.333,.337,.335,.330,.328,.332,.330,.331,.333
930 DATA .330,.330,.334,.338,.327,.330,.327,.331,.330,.329
940 DATA .329,.332,.331,.330,.328,.328,.328,.328,.336,.334
950 DATA .329,.329,.332,.327,.327,.333,.337,.333,.325,.324
960 DATA .328,.331,.330,.332,.335,.332,.332,.325,.329,.331
970 DATA .334,.329,.327,.334,.331,.332,.332,.327,.337,.332
980 DATA .328,.331,.327,.332,.331,.330,.335,.330,.332,.332
990 DATA 4,25,50,75,95
1000 END
```

Output for data of Table 2.1

```
DISTRIBUTION MEAN IS        .33142
DISTRIBUTION VARIANCE IS  9.97333313E-06
DISTRIBUTION S.D. IS      3.15805844E-03
```

CLASS MID-VALUE	FREQUENCY [†]
.3245	3
.3265	7
.3285	18
.3305	23
.3325	24
.3345	15
.3365	7
.3385	3

```
25 TH PERCENTILE IS  .329166667
50 TH PERCENTILE IS  .331413043
75 TH PERCENTILE IS  .3335
95 TH PERCENTILE IS  .336928571
```

† To obtain the tabular format above (and in subsequent programs) extra programming statements, which are not included in the program listings, are required.

(c) Program 2.3

```
100 PRINT"□":REM**CLEAR SCREEN
110 PRINT"           **PROGRAM 2.3**"
120 PRINT"           STEM AND LEAF PLOT"
130 PRINT:PRINT
140 DIM L(20,30),F(20)
150 REM**READ SAMPLE SIZE N,NUMBER OF CLASSES NC
160 REM**LOWER BOUNDARY B OF FIRST CLASS,CLASS WIDTH W
170 REM**LOWEST POWER PM OF 10 SUCH THAT
180 REM**10↑PM*X(I) IS INTEGRAL FOR ALL X(I)
190 READ N,NC,B,W,PM
200 REM**PUT LEAVES INTO ARRAYS L(I,J)(I=1 TO NC,J=1 TO F(I))
210 REM**F(I) IS THE FREQUENCY IN CLASS I
220 REM**INITIALISE ARRAYS TO ZERO
230 FOR I=1 TO NC:F(I)=0:FOR J=1 TO 30:L(I,J)=0:NEXT J:NEXT I
240 REM**READ IN AND PRINT OUT DATA
250 PRINT"SAMPLE VALUES ARE"
260 FOR I=1 TO N
270 READ XI
280 PRINT XI;
290 IF I=INT(I/10)*10 THEN PRINT
300 REM**DETERMINE CLASS J OF XI
310 J=INT((XI-B)/W)+1
320 F(J)=F(J)+1
330 L(J,F(J))=XI
340 NEXT I
350 PRINT
360 PRINT"PRESS ANY KEY TO CONTINUE"
370 GET A$ : IF A$="" THEN 370
380 PRINT"□":REM**CLEAR SCREEN
390 PRINT"   STEM AND LEAF PLOT"
400 PRINT
```

```
410 REM**ORDERING OF LEAVES
420 FOR I=1 TO NC
430 IF F(I)<2 THEN 510
440 FOR J=1 TO F(I)
450 FOR J1=J TO F(I)
460 IF L(I,J1)>=L(I,J) THEN 480
470 TEMP=L(I,J) : L(I,J)=L(I,J1) : L(I,J1)=TEMP
480 REM
490 NEXT J1
500 NEXT J
510 REM
520 NEXT I
530 REM**CALCULATION OF LOWEST VALUE XL IN FIRST CLASS
540 XL=B+0.5/10↑PM
550 FOR I=1 TO NC
560 REM**CALCULATION OF STEM XST FOR CLASS I
570 XST=INT(XL*10↑(PM-1)+1E-05)/10↑(PM-1)
580 A$=STR$(XST)+" "
590 IF F(I)=0 THEN 680
600 REM**ADDITION OF LEAVES Y TO STEM
610 FOR K=1 TO F(I)
620 Y=INT(10↑PM*(L(I,K)-XST)+1E-05)
630 IF Y<10 THEN NS=1
640 IF Y>9 THEN NS=2
650 A$=A$+MID$(STR$(Y),2,NS)
660 IF K=INT(K/5)*5 AND K<F(I) THEN A$=A$+" "
670 NEXT K
680 PRINT A$
690 A$="" : XL=XL+W
700 NEXT I
710 PRINT
720 REM**DATA FROM TABLE 2.1
730 DATA 100,8,0.3235,0.002,3
740 DATA .339,.329,.337,.330,.329,.330,.338,.335,.331,.336
750 DATA .330,.333,.329,.337,.334,.328,.334,.333,.329,.333
760 DATA .332,.334,.334,.330,.335,.332,.335,.332,.333,.335
770 DATA .333,.333,.337,.335,.330,.328,.332,.330,.331,.333
780 DATA .330,.330,.334,.338,.327,.330,.327,.331,.330,.329
790 DATA .329,.332,.331,.330,.328,.328,.328,.328,.336,.334
800 DATA .329,.329,.332,.327,.327,.333,.337,.333,.325,.324
810 DATA .328,.331,.330,.332,.335,.332,.332,.325,.329,.331
820 DATA .334,.329,.327,.334,.331,.332,.332,.327,.337,.332
830 DATA .328,.331,.327,.332,.331,.330,.335,.330,.332,.332
850 END
READY.
```

Stem and leaf plot for data of Table 2.1

```
        STEM AND LEAF PLOT

  .32 455
  .32 77777 77
  .32 88888 88899 99999 999
  .33 00000 00000 00001 11111 111
  .33 22222 22222 22222 33333 3333
  .33 44444 44455 55555
  .33 66777 77
  .33 889
```

APPENDIX 2.1 DERIVATION OF ALTERNATIVE FORMULA FOR SAMPLE VARIANCE

$$s^2 = \frac{1}{n-1} \sum_{i=1}^{k} f_i(x_i - \bar{x})^2$$

$$= \frac{1}{n-1} \sum_{i=1}^{k} (f_i x_i^2 - 2f_i x_i \bar{x} + f_i \bar{x}^2)$$

$$= \frac{1}{n-1} \left(\sum_{i=1}^{k} f_i x_i^2 - 2\bar{x} \sum_{i=1}^{k} f_i x_i + \bar{x}^2 \sum_{i=1}^{k} f_i \right)$$

$$= \frac{1}{n-1} \left(\sum_{i=1}^{k} f_i x_i^2 - 2\bar{x} n \bar{x} + \bar{x}^2 n \right)$$

using $\displaystyle\sum_{i=1}^{k} f_i x_i = n\bar{x}$ and $\displaystyle\sum_{i=1}^{k} f_i = n$

$$= \frac{1}{n-1} \left(\sum_{i=1}^{k} f_i x_i^2 - n\bar{x}^2 \right)$$

$$= \frac{1}{n-1} \left\{ \sum_{i=1}^{k} f_i x_i^2 - \frac{1}{n} \left(\sum_{i=1}^{k} f_i x_i \right)^2 \right\} \tag{2.7}$$

using $\displaystyle\bar{x} = \frac{1}{n} \sum_{i=1}^{k} f_i x_i$.

Putting $f_i = 1, i = 1, 2, \ldots, k$, gives $k = n$ and

$$s^2 = \frac{1}{n-1} \left\{ \sum_{i=1}^{n} x_i^2 - \frac{1}{n} \left(\sum_{i=1}^{n} x_i \right)^2 \right\} . \tag{2.6}$$

APPENDIX 2.2 CODING

Let $x_i = cu_i + \alpha$.

Then $\bar{x} = \dfrac{1}{n} \sum\limits_{i=1}^{k} f_i x_i$

$$= \frac{1}{n} \sum_{i=1}^{k} f_i (cu_i + \alpha)$$

$$= c \frac{1}{n} \sum_{i=1}^{k} f_i u_i + \alpha \frac{1}{n} \sum_{i=1}^{k} f_i$$

$$= c\bar{u} + \alpha \qquad \text{since } \sum_{i=1}^{k} f_i = n \qquad . \tag{2.9}$$

$$s_x^2 = \frac{1}{n-1} \sum_{i=1}^{k} f_i (x_i - \bar{x})^2$$

$$= \frac{1}{n-1} \sum_{i=1}^{k} f_i (cu_i + \alpha - (c\bar{u} + \alpha))^2, \quad \text{using (2.8) and (2.9)}$$

$$= c^2 \frac{1}{n-1} \sum_{i=1}^{k} f_i (u_i - \bar{u})^2$$

$$= c^2 s_u^2 \qquad . \tag{2.10}$$

EXERCISES

Section A

1. The weights of a random sample of ten sodium bicarbonate tablets are (in grams)

   0.469   0.442   0.446   0.416   0.444   0.366   0.430   0.425   0.400   0.445

   Find the mean, variance and standard deviation of the sample.

2. In an experiment to examine the effect of glucagon on blood sugar levels the following blood sugar concentrations (in mg/100 ml blood) are found in a group of eight rats 15 minutes after dosage with glucagon

   270   265   280   265   275   250   245   260

   Find the mean, variance and standard deviation of the sample.

3. The weights of a random sample of twenty aspirin tablets are (in grams)

   0.332   0.333   0.334   0.336   0.329   0.328   0.336   0.335   0.331   0.335
   0.328   0.328   0.334   0.326   0.333   0.332   0.328   0.327   0.328   0.332

   Find the mean, variance and standard deviation of the sample.

4. The weights of a sample of 30 capsules are (to the nearest 5 mg)

   360,   340,   350,   350,   355,   345,   345,   345,   355,   355,   360,   350,
   350,   350,   365,   350,   335,   340,   355,   355,   345,   350,   360,   355,
   355,   350,   340,   345,   350,   355.

   Reduce the sample to a frequency distribution and find the mean, variance and standard deviation of the distribution. Illustrate the distribution by a histogram.

5. The blood alcohol concentration required to cause respiratory failure in a sample of 233 rats is tabulated below

Class mid-value	8.8	8.9	9.0	9.1	9.2	9.3	9.4	9.5	9.6	9.7
Frequency	1	2	17	20	43	62	56	23	7	2

   Find the mean, variance and standard deviation of the distribution. Draw a histogram of the results.

6. Water is run from a burette up to the 10 ml mark in a measure by each of the members of a class and the differences between the initial and

final burette readings obtained. The results (in ml) obtained by the class are tabulated below

10.0	10.3	9.2	8.7	10.2	8.8	9.6	9.7	10.1	10.1
9.6	10.3	9.6	9.7	9.9	9.3	9.2	9.8	9.8	9.2
9.8	9.0	9.7	9.9	10.1	9.9	9.4	9.8	10.0	10.1
10.2	10.2	9.9	9.9	10.2	9.8	9.8	10.2	10.1	9.5
10.0	10.3	10.1	10.3	10.2	9.6	9.9	9.9	10.1	10.1
10.1	10.0	10.0	10.3						

Reduce the results to a frequency distribution and find the mean, variance and standard deviation of the distribution. Draw a histogram of the distribution.

7. The weights of a hundred sodium bicarbonate tablets are tabulated below (in grams)

0.485	0.469	0.442	0.446	0.416	0.444	0.366	0.430	0.425	0.400
0.445	0.388	0.429	0.427	0.402	0.441	0.438	0.444	0.419	0.443
0.448	0.433	0.412	0.436	0.414	0.370	0.445	0.441	0.454	0.424
0.407	0.439	0.435	0.450	0.394	0.416	0.367	0.383	0.425	0.376
0.458	0.388	0.390	0.379	0.426	0.446	0.391	0.410	0.419	0.440
0.391	0.362	0.422	0.439	0.466	0.405	0.412	0.434	0.445	0.414
0.443	0.440	0.362	0.442	0.366	0.446	0.453	0.478	0.445	0.425
0.419	0.408	0.452	0.450	0.464	0.440	0.458	0.448	0.443	0.438
0.453	0.415	0.391	0.423	0.427	0.390	0.327	0.422	0.419	0.426
0.413	0.410	0.461	0.413	0.414	0.428	0.417	0.392	0.464	0.424

Reduce the results to a frequency distribution with classes 0.3195–0.3395, 0.3395–0.3595, etc. Find the mean, variance and standard deviation of the distribution. Draw a histogram of the distribution and comment on the form of the distribution.

8. Find the medians and quartiles of the samples of examples 1, 2 and 3.

9. Find the median, 10th and 90th percentiles of the data of example 4.

10. Find the medians, quartiles and 5th and 95th percentiles of the distributions in examples 5 and 7. Find also the semi-interquartile range in each case.

11. Draw box-and-dot plots and stem-and-leaf plots for the distributions in examples 6 and 7.

12. A sample of 10 tablets has a mean weight of 0.4323 g with a variance of 1.1451×10^{-3} g^2; a further sample of 5 tablets has a mean weight of 0.4182 g with a variance of 5.2170×10^{-4} g^2. Find the mean and variance of the combined sample of 15 tablets.

13. Check the calculation of the means and standard deviations of the data in examples 3 and 5 by first coding the data.

Section B

1. The following table shows the variation of the coefficient of friction μ with the angle of apex of the sliding cone for various metals

Cone angle (degrees)	Al	Pb	Mild Steel	Stainless steel	Brass
60	0.84	0.85	0.52	0.52	0.52
90	0.73	0.75	0.50	0.34	0.44
105	0.97	0.75	0.50	0.51	0.61
120	0.94	0.73	0.46	0.48	0.58
136	0.80	0.64	0.40	0.45	0.45
150	0.63	0.48	0.35	0.33	0.31
170	0.22	0.18	0.20	0.15	0.18

Find (i) the means and standard deviations of the values of μ obtained for each cone angle,
   (ii) the means and standard deviations of the values of μ obtained for each material,
   (iii) the mean and standard deviation of all the values of μ.

2. The breaking strengths of 100 steel specimens are given in the following frequency table

Centre of interval	32.0	32.2	32.4	32.6	32.8	33.0	33.2	33.4	33.6	33.8
Frequency	1	4	11	13	21	24	12	9	3	2

Find the mean and standard deviation of these results.

3. The speeds of 50 vehicles in kph are

65	41	63	53	30	58	19	87	85	109
77	47	66	54	59	78	87	39	80	43
60	85	53	68	89	73	27	79	63	82
71	46	74	49	26	56	55	74	54	50
110	62	38	17	98	64	25	42	82	118

Reduce these results to a frequency distribution with intervals 9.5–19.5 etc., and calculate the mean and standard deviation of the distribution. Draw a histogram of the distribution.

4. The breaking strengths of 100 specimens of a material are given in the following table

7.16	7.12	6.63	6.85	6.89	7.08	7.01	7.14	6.71	6.95
7.26	7.01	6.99	6.64	6.81	7.19	7.24	6.80	6.90	7.00
7.42	6.91	6.92	7.24	6.97	6.92	7.04	6.77	6.56	6.76
7.15	7.07	6.95	7.14	6.96	7.45	7.25	7.21	7.28	6.94
6.99	7.06	7.03	6.92	7.07	7.16	6.85	7.09	6.92	6.82
6.60	6.88	7.14	6.99	7.17	6.94	6.69	6.97	6.88	7.10
6.84	6.91	7.38	7.26	6.82	6.73	6.91	6.96	6.96	6.98
7.22	6.93	6.76	7.03	7.17	7.05	6.54	7.10	7.36	7.08
7.01	7.05	6.62	6.87	7.05	7.33	6.80	7.02	7.24	6.96
6.90	7.28	7.40	6.98	6.94	7.12	6.79	6.89	6.85	7.02

Reduce the results to a frequency distribution with classes 6.495–6.595, 6.595–6.695, ... Find the mean and standard deviation of the distribution.

5. Find the medians and quartiles of the values of μ for each of the metals in example 1.

6. Find the median, 10th and 90th percentiles of the distribution of example 4.

7. Draw box and dot plots and stem and leaf plots for the distributions of examples 3 and 4.

8. A sample of 10 observations of SO_2 concentration at a site have a mean of 78.6 $\mu g/m^3$ with a standard deviation of 10.4690; a further sample of 5 observations has a mean of 74.6 $\mu g/m^3$ with a standard deviation of 10.2127. Find the mean and variance of the combined sample of 15 observations.

9. Check the calculation of the means and standard deviations in examples 2 and 3 by first coding the data.

Section C†

1. Use programs 2.1, 2.2 and 2.3 to check where appropriate the answers to the examples of sections A and B.

2. Write a program to evaluate the mean, standard deviation and percentiles of a sample which has already been reduced to a frequency distribution.

3. Write a program for coding data, evaluating the mean and standard deviation of the coded data, and obtaining from these the statistics of the original data.

†Check the operation of programs developed in this section on data from the examples of sections A and B, or worked examples in the text.

4. Write a program to reduce a sample of discrete observations to a frequency distribution and to evaluate the sample statistics.

5. Write a program to evaluate the sample statistics necessary for drawing a box and dot plot.

6. Adapt program 2.1 (i) to accept data from the keyboard, (ii) to print out the data to the screen, (iii) to ask whether the data are correct (iv) to accept corrections if necessary and (v) to evaluate the mean and standard deviation of the data.

3

Elementary Probability Theory

3.1 INTRODUCTION

Consider an experiment whose outcome is random; that is, before the experiment is performed, we know that its outcome will be one of a set of possible outcomes, but it is impossible to predict which one with certainty. The *sample space S* associated with the experiment is the collection of all possible outcomes of the experiment. Any particular outcome is called an *event*. An event which cannot be broken down into a combination of simpler events is called an *elementary event*. These concepts are illustrated in the examples described below.

Experiment 1. Tossing a coin. There are two outcomes (events) associated with this experiment, obtaining a head (H), or a tail (T). These are also elementary events and together constitute the sample space for the experiment. They are said to be *mutually exclusive* as they cannot occur as the result of the same trial of the experiment, and *exhaustive* because there are no other elementary events.

Experiment 2. Throwing three coins simultaneously. The eight elementary, mutually exclusive and exhaustive elementary events associated with this experiment are *HHH, HHT, HTH, THH, HTT, THT, TTH, TTT* and these constitute the sample space. Examples of events which are not elementary are E_1, obtaining exactly two heads, and E_2 obtaining one or more tails.

Experiment 3. Counting the number of disintegrations from a radioactive source in a given interval of time. The sample space is $X = 0, 1, 2, 3, \ldots$ disintegrations. $X = 3$ is an example of an elementary event and $X \geqslant 3$ of an event which is not elementary.

Experiment 4. Measuring the yield Y g of a product in a chemical reaction. The sample space is $Y \geqslant 0$. An example of an event is $10 \leqslant Y \leqslant 20$.

Experiment 5. Weighing four aspirin tablets and finding their mean weight $\bar{X}$ g. The sample space is $\bar{X} \geqslant 0$. $\bar{X} > 0.300$ is an example of an event.

In statistical analysis it is frequently necessary to evaluate the probabilities of events. In order to do this in some situations, the event concerned has to be expressed as a combination of simpler events. The rules for the evaluation of the probabilities of such events will be discussed in section 3.3.

3.2 FREQUENCY DEFINITION OF PROBABILITY

Let E be an event associated with the outcome of an experiment. Suppose that the experiment is repeated N times and that the event E occurs in n of these repetitions. Let $p = \Pr(E)$ be the probability of the event E. Then p is defined by

$$p = \lim_{N \to \infty} \frac{n}{N} . \tag{3.1}$$

In words, p is the limiting value of the ratio n/N as N becomes very large. For convenience we shall write (3.1) in the form

$$p = \frac{n(L)}{N(L)} \tag{3.2}$$

to remind us that N and n are very Large. The use of this notation will simplify the explanation of the results considered in the following sections.

The manner in which p approaches its limiting value is illustrated in Fig. 3.1 for the results of a computer simulation of tossing a fair coin. Here the event E is

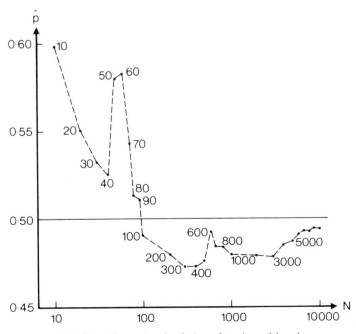

Fig. 3.1 – Computer simulation of tossing a fair coin.

obtaining a head; because the coin is assumed to be fair, the theoretical value of the probability of E is $1/2$.

It will be observed that a considerable number of repetitions is required in order to obtain a reasonably close approximation to the theoretical value of the probability, even for such a simple experiment as tossing a coin. Fortunately it is not very often necessary in practice to estimate probabilities in this way in order to make statistical inferences!

If only N repetitions of an experiment are carried out, the estimate $\hat{p}$ of p is simply

$$\hat{p} = \frac{n}{N} \ . \tag{3.3}$$

Suppose that the event E_0 is impossible; then E_0 cannot occur as an outcome of any of the repetitions of the experiment so that $n = 0$ for all N. Then

$$\Pr(E_0) = \frac{0}{N(L)} = 0 \ . \tag{3.4}$$

Suppose that the event S is certain; then S occurs at each repetition so that $n = N$ for all N and

$$\Pr(S) = \frac{N(L)}{N(L)} = 1 \ . \tag{3.5}$$

As the probability p of any event must lie between these two extremes, then in general

$$0 \leqslant p \leqslant 1 \ . \tag{3.6}$$

Let $\bar{E}$ be the event that E does not occur. Then if E occurs in n out of the N repetitions of the experiment, $\bar{E}$ must occur in the remaining $N-n$ repetitions. Hence

$$\Pr(E) + \Pr(\bar{E}) = \frac{n(L)}{N(L)} + \frac{N(L) - n(L)}{N(L)}$$

$$= \frac{n(L) + N(L) - n(L)}{N(L)}$$

$$= 1$$

and thus $\quad \Pr(\bar{E}) = 1 - \Pr(E) \ . \tag{3.7}$

3.3 PROBABILITIES OF COMBINATIONS OF EVENTS

Let E_1 and E_2 be any two events. Then

the event '$E_1 E_2$' occurs if E_1 *and* E_2 have occurred

and

the event '$E_1 + E_2$' occurs if *either* E_1 *or* E_2 or *both* have occurred.

We now examine methods of evaluating the probabilities of the events $E_1 E_2$ and $E_1 + E_2$.

(a) Calculation of the probability of the event $E_1 E_2$.

We first introduce the idea of the *conditional probability* of E_1 given that E_2 has occurred (written as $\Pr(E_1 | E_2)$). Suppose that in N repetitions of an experiment the event E_1 occurs n_1 times, the event E_2 occurs n_2 times, and the event $E_1 E_2$ occurs n_{12} times. Thus on the n_2 occasions on which E_2 occurs, E_1 also occurs n_{12} times. Hence from the frequency distribution of probability

$$\Pr(E_1) = \frac{n_1(L)}{N(L)}, \Pr(E_2) = \frac{n_2(L)}{N(L)}, \Pr(E_1 | E_2) = \frac{n_{12}(L)}{n_2(L)} \text{ and } \Pr(E_1 E_2) = \frac{n_{12}(L)}{N(L)},$$

where L (as before) implies that the experiment has been repeated a large number of times. Then

$$\Pr(E_1 | E_2) = \frac{n_{12}(L)}{n_2(L)} = \frac{n_{12}(L)/N(L)}{n_2(L)/N(L)}$$

(dividing the numerator and denominator by $N(L)$)

$$= \frac{\Pr(E_1 E_2)}{\Pr(E_2)} . \tag{3.8}$$

By similar reasoning, or simply from symmetry, it is seen that

$$\Pr(E_2 | E_1) = \frac{\Pr(E_1 E_2)}{\Pr(E_1)} . \tag{3.9}$$

Two events E_1 and E_2 are said to be *independent* if the probability of E_1 occurring is the same whether or not E_2 has occurred. Thus for independent events

$$\Pr(E_1 | E_2) = \Pr(E_1) \tag{3.10}$$

Thus from (3.8) when E_1 and E_2 are independent

$$\Pr(E_1 E_2) = \Pr(E_1) \times \Pr(E_2) \tag{3.11}$$

and from (3.9) and (3.11)

$$\Pr(E_2|E_1) = \Pr(E_2) \ . \tag{3.12}$$

Similarly if the events E_1, E_2 and E_3 are mutually independent

$$\Pr(E_1 E_2 E_3) = \Pr(E_1) \times \Pr(E_2) \times \Pr(E_3) \ . \tag{3.13}$$

(b) Calculation of the probability of the event $E_1 + E_2$
The first case considered is when E_1 and E_2 are mutually exclusive (that is they cannot occur together). Suppose that out of N repetitions of an experiment E_1 occurs n_1 times and E_2 occurs n_2 times. Since E_1 and E_2 are mutually exclusive $E_1 E_2$ does not occur. Hence $E_1 + E_2$ (which by definition is either E_1 or E_2 or both) occurs $n_1 + n_2$ times. Then

$$\Pr(E_1 + E_2) = \frac{n_1(L) + n_2(L)}{N(L)} = \frac{n_1(L)}{N(L)} + \frac{n_2(L)}{N(L)} = \Pr(E_1) + \Pr(E_2) \ .$$
$$\tag{3.14}$$

Similarly if the events E_1, E_2 and E_3 are mutually exclusive then

$$\Pr(E_1 + E_2 + E_3) = \Pr(E_1) + \Pr(E_2) + \Pr(E_3) \ . \tag{3.15}$$

Formula (3.14) may be adapted for events E_1 and E_2 which are not mutually exclusive as follows. As before let n_{12} be the number of times that the event $E_1 E_2$ occurs in N repetitions of the experiment. It is now incorrect to say that the event $E_1 + E_2$ occurs $n_1 + n_2$ times (as for the mutually exclusive case) since the n_{12} occurrences of $E_1 E_2$ would then be counted twice; to correct for this we must take n_{12} away to give $n_1 + n_2 - n_{12}$ occurrences for the event $E_1 + E_2$. Thus

$$\Pr(E_1 + E_2) = \frac{n_1(L) + n_2(L) - n_{12}(L)}{N(L)} = \frac{n_1(L)}{N(L)} + \frac{n_2(L)}{N(L)} - \frac{n_{12}(L)}{N(L)}$$

$$= \Pr(E_1) + \Pr(E_2) - \Pr(E_1 E_2) \ . \tag{3.16}$$

The above result may be extended to evaluate $\Pr(E_1 + E_2 + E_3)$ when the events E_1, E_2 and E_3 are not mutually exclusive, but this extension will not be considered here.

It will be noted that (3.14) is a particular case of (3.16) when $\Pr(E_1 E_2) = 0$, corresponding to E_1 and E_2 being mutually exclusive.

The application of the above results is illustrated in the examples below.

Ex. 3.1. The operation of a switch S_1 is illustrated in the diagram. The switch is in positions 1, 2 and 3 with probabilities p_1, p_2 and p_3 respectively (with $p_1 + p_2 + p_3 = 1$). Find the probability that current can flow between the points A and B.

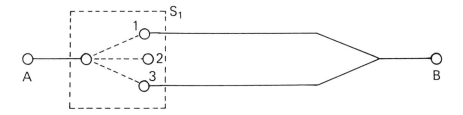

Let C_i be the event that the switch is in position i ($i = 1, 2, 3$). Then C_1, C_2 and C_3 are mutually exclusive and exhaustive. Thus the sample space S for the experiment can be written as $S = C_1 + C_2 + C_3$ and

$$1 = \Pr(S) = \Pr(C_1 + C_2 + C_3) = \Pr(C_1) + \Pr(C_2) + \Pr(C_3) = p_1 + p_2 + p_3$$

Let F be the event that current can flow from A to B. Then

$$\Pr(F) = \Pr(C_1 + C_3)$$
$$= \Pr(C_1) + \Pr(C_3) \text{ from (3.14), since } C_1 \text{ and } C_3 \text{ are}$$
$$\text{mutually exclusive}$$
$$= p_1 + p_3 \ .$$

Alternatively

$$\Pr(F) = \Pr(\bar{C}_2)$$
$$= 1 - \Pr(C_2) \qquad \text{from (3.7)}$$
$$= 1 - p_2$$
$$= p_1 + p_3 \ .$$

Ex. 3.2. Two switches S_1 and S_2 are connected in series. Let C_i be the event that S_i is closed ($i = 1, 2$). If C_1 and C_2 are independent with probabilities p_1 and p_2 respectively find the probability that current can flow between A and B.

$$\Pr(F) = \Pr(C_1 C_2)$$
$$= \Pr(C_1) \times \Pr(C_2) \text{ from (3.11), since } C_1 \text{ and } C_2 \text{ are}$$
$$\text{independent}$$
$$= p_1 p_2$$

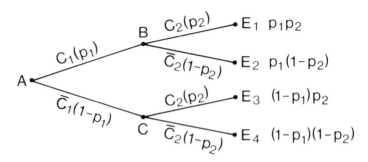

Fig. 3.2 – Tree diagram for examples 3.2 and 3.3.

The outcomes of this experiment are illustrated in the *tree diagram* of Fig. 3.2. The branches AB and AC correspond to the two possible states of the switch S_1 (C_1 with probability p_1 and $\bar{C}_1$ with probability $(1 - p_1)$). Corresponding to the state C_1 of S_1 there are two possible states of S_2, namely C_2 with probability p_2 and $\bar{C}_2$ with probability $(1 - p_2)$ (if the states of S_1 and S_2 were not independent the conditional probabilities $\Pr(C_2|C_1)$ and $\Pr(\bar{C}_2|C_1)$ would have been used here); these states correspond to the branches BE_1 and BE_2. In a similar manner the branches CE_3 and CE_4 are constructed. The events E_1, E_2, E_3 and E_4 form the set of mutually exclusive elementary events for the outcomes of this experiment. The probability of the event E_i may be found by multiplying together the probabilities on the branches leading to the event. Then

$$\Pr(F) \;=\; \Pr(E_1) = p_1 p_2 \quad \text{as above.}$$

Note that the probability that no current can flow between A and B, $\Pr(\bar{F})$, may be calculated either as $\Pr(E_2 + E_3 + E_4)$, or, more simply, as $\Pr(\bar{E}_1)$ $= 1 - \Pr(E_1)$.

Ex. 3.3. If the two switches of the previous example are connected in parallel as illustrated in the diagram, find the probability that current can flow between A and B.

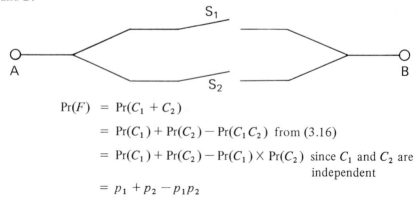

$$\Pr(F) \;=\; \Pr(C_1 + C_2)$$

$$= \Pr(C_1) + \Pr(C_2) - \Pr(C_1 C_2) \quad \text{from (3.16)}$$

$$= \Pr(C_1) + \Pr(C_2) - \Pr(C_1) \times \Pr(C_2) \quad \text{since } C_1 \text{ and } C_2 \text{ are}$$
$$\text{independent}$$

$$= p_1 + p_2 - p_1 p_2$$

Alternatively the required probability may be evaluated by reference to the tree diagram for the experiment, which is the same as that for the previous example (Fig. 3.2); thus

$$\Pr(F) = \Pr(E_1) + \Pr(E_2) + \Pr(E_3)$$
$$= p_1 p_2 + p_1(1 - p_2) + (1 - p_1)p_2$$
$$= p_1 + p_2 - p_1 p_2 \text{ as above .}$$

Ex. 3.4. In the situation of example 3.3 find the probability (a) that switch S_1 is closed given that a current can flow from A to B, (b) that a current can flow from A to B given that switch S_1 is closed.

(a) $\qquad \Pr(C_1|F) = \dfrac{\Pr(C_1 F)}{\Pr(F)}$ from (3.9)

$$= \frac{\Pr(E_1 + E_2)}{\Pr(E_1 + E_2 + E_3)} \quad \text{from Fig. 3.2}$$

$$= \frac{\Pr(E_1) + \Pr(E_2)}{\Pr(E_1) + \Pr(E_2) + \Pr(E_3)}$$

$$= \frac{p_1 p_2 + p_1(1 - p_2)}{p_1 p_2 + p_1(1 - p_2) + (1 - p_1)p_2}$$

$$= \frac{p_1}{p_1 + p_2 - p_1 p_2} \ .$$

(b) $\qquad \Pr(F|C_1) = \dfrac{\Pr(FC_1)}{\Pr(C_1)}$ from (3.9)

$$= \frac{p_1}{p_1} \qquad \text{using } \Pr(FC_1) = p_1 \text{ from (a) above}$$

$$= 1 \ .$$

(In retrospect it is obvious that a current is certain to flow from A to B if S_1 is closed!)

Ex. 3.5. A box contains 3 defective and 7 satisfactory fuses. Two fuses are chosen at random and without replacement from the box. Find the probabilities that (a) both fuses are defective, (b) the first fuse is defective and the second is not, (c) exactly one of the fuses is defective.

(a) If the first fuse is defective (D), there will be 2 defective and 7 satisfactory (S) fuses left in the box; then

$$\Pr(D_2|D_1) = \frac{2}{9} \;.$$

Hence $\Pr(D_2 D_1) = \Pr(D_2|D_1) \times \Pr(D_1)$ from (3.9)

$$= \frac{2}{9} \times \frac{3}{10}$$

$$= \frac{6}{90} = \frac{1}{15} \;.$$

(b) $\Pr(S_2 D_1) = \Pr(S_2|D_1) \times \Pr(D_1)$

$$= \frac{7}{9} \times \frac{3}{10}$$

$$= \frac{21}{90} = \frac{7}{30} \;.$$

(c) Let X be the number of defective fuses in the sample. Then

$$\Pr(X = 1) = \Pr(D_2 S_1 + S_2 D_1)$$

$$= \Pr(D_2 S_1) + \Pr(S_2 D_1) \quad \text{since } D_2 S_1 \text{ and } S_2 D_1 \text{ are}$$
$$\text{mutually exlusive}$$

$$= \frac{3}{9} \times \frac{7}{10} + \frac{7}{9} \times \frac{3}{10}$$

$$= \frac{42}{90} = \frac{7}{15} \;.$$

As in previous examples an alternative approach to the evaluation of the probabilities is by reference to a tree diagram (Fig. 3.3).

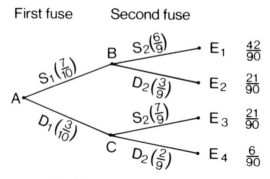

Fig. 3.3 – Tree diagram for example 3.5.

It is left as an exercise to the reader to verify the answers obtained above by reference to the diagram.

3.4 BAYES' THEOREM

Let $E_1, E_2, \ldots, E_k$ be a set of mutually exclusive exhaustive events and let A, which is dependent on the E_is, be an outcome of an experiment. Then

$$\Pr(E_i|A) = \frac{\Pr(A|E_i) \times \Pr(E_i)}{\Pr(A|E_1)\Pr(E_1) + \ldots + \Pr(A|E_k)\Pr(E_k)} \qquad i = 1. \ldots, k \ . \quad (3.17)$$

This result is known as Bayes' theorem. In particular when $k = 2$ Bayes' theorem becomes

$$\Pr(E_i|A) = \frac{\Pr(A|E_i)\Pr(E_i)}{\Pr(A|E_1)\Pr(E_1) + \Pr(A|E_2)\Pr(E_2)} \qquad i = 1, 2 \ . \quad (3.18)$$

A derivation of (3.18) may be found in Appendix 3.1.

The application of Bayes' theorem is illustrated in the example below.

Ex. 3.6. The probability of a person chosen at random from a population having a particular disease is 0.02. The probability that a diagnostic test gives a positive result when the disease is present is 0.75 and a negative result when the disease is not present is 0.97.

What is the probability of a person having the disease (a) when the test gives a positive result and (b) when the test gives a negative result?

Let D be the event that the person has the disease (and $\bar{D}$ the event that he does not).

Let P be the event that the test is positive (and $\bar{P}$ the event that it is negative).

Then we are asked to find (a) $\Pr(D|P)$ and (b) $\Pr(D|\bar{P})$.

From the information in the question

$$\Pr(P|D) = 0.75 \quad \text{(and hence } \Pr(\bar{P}|D) = 1 - \Pr(P|D) = 0.25)$$
$$\Pr(\bar{P}|\bar{D}) = 0.97 \quad \text{(and hence } \Pr(P|\bar{D}) = 1 - \Pr(\bar{P}|\bar{D}) = 0.03)$$
$$\Pr(D) = 0.02 \quad \text{(and hence } \Pr(\bar{D}) = 1 - \Pr(D) = 0.98)$$

From Bayes' theorem with $k = 2$ (3.18)

$$\Pr(D|P) = \frac{\Pr(P|D)\Pr(D)}{\Pr(P|D)\Pr(D) + \Pr(P|\bar{D})\Pr(\bar{D})}$$

$$= \frac{0.75 \times 0.02}{0.75 \times 0.02 + 0.03 \times 0.98}$$

$$= 0.338$$

and

$$Pr(D|\bar{P}) = \frac{Pr(\bar{P}|D)\,Pr(D)}{Pr(\bar{P}|D)\,Pr(D) + Pr(\bar{P}|\bar{D})\,Pr(\bar{D})}$$

$$= \frac{0.25 \times 0.02}{0.25 \times 0.02 + 0.97 \times 0.98}$$

$$= 0.0052 \ .$$

3.5 COMPUTER SIMULATION OF SIMPLE RANDOM EXPERIMENTS

In this section computer programs will be introduced to *simulate* some of the experimental situations described earlier in the chapter.

Suppose that $E_1, \ldots, E_k$ are a set of mutually exclusive exhaustive events with

$$p_i = Pr(E_i) \qquad i = 1, 2, \ldots, k$$

so that

$$p_1 + p_2 + \ldots + p_k = 1 \ .$$

A computer can be used to generate a random number U in the range 0 to 1. If the number falls in the range $0 \leqslant U \leqslant p_1$ (Fig. 3.4), we can consider the event E_1 to have occurred; if it falls in the range $p_1 < U \leqslant p_2$ we can consider the event E_2 to have occurred, and so on. To simulate the action of a switch, for example, which is closed with probability p, the following procedure can be used. Let E_1 be the event that the switch is closed $(Pr(E_1) = p)$ and E_2 the event that the switch is open $(Pr(E_2) = 1 - p)$. Generate a random number U lying in the range 0 to 1; if $0 \leqslant U \leqslant p$ take the switch to be closed; if $p < U \leqslant 1$ take the switch to be open.

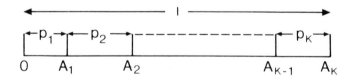

Fig. 3.4 – Computer simulation of a set of events with given probabilities.

The instruction U = RND(1) on the COMMODORE computer returns random numbers U distributed uniformly over the range 0 to 1. To repeat the same sequence of numbers, the statement X0 = RND(−N), were −N is a fixed negative integer, may be used to initiate the sequence.

Strictly speaking the sequences produced by the computer are pseudo-random numbers rather than purely random numbers since, as mentioned above, the same sequence may be reproduced as often as wished by using a particular starting 'seed' $-N$ for the generating algorithm. One such algorithm is based on the Lehmer-congruence method. This method uses two constants k and m and produces a sequence of pseudo-random numbers $x_1, x_2, \ldots, x_r, x_{r+1} \ldots$ as follows. Multiply x_r by k, divide by m and put x_{r+1} equal to the remainder. The properties of the sequence obtained will depend on the values of k and m used; for example, certain pairs of values of k and m can lead to sequences which repeat themselves after an unacceptably small interval. Sequences can be tested for their cyclic properties, for correlation, for the relative frequencies of occurrence of different combinations of digits, and for the randomness of the interval between occurrences of the same digit or same sequence of digits. A further discussion of random number generation may be found for example in Oldknow and Smith (1983).

Program 3.1 simulates the experiment of tossing a coin, where the probability p of getting heads is set by the user. Typical output from this program (with $p = 0.5$) leads to the results illustrated in Fig. 3.1. The program also prints out the numbers of the tosses at which the estimate of p coincides with its true value.

Program 3.2 simulates the switching experiments examined in examples 3.2, 3.3 and 3.4. Here again the switching probabilities are set by the user.

It will be observed from the output for both programs that very large numbers of repetitions are required before the probability estimates approximate the true values to any degree of accuracy. It will also be noted that the probability estimates tend to remain above or below the true values for long periods. The general nature of these aspects of the experiments can be verified by running the programs using different sequences of random numbers and different numbers of repetitions.

Program 3.3 simulates the illustration of Bayes' theorem in example 3.6. The current number N of persons is partitioned as follows (where D – diseased; H (or $\bar{D}$) – healthy; P – positive test; N (or $\bar{P}$) – negative test):

	D	H	
P	n_{DP}	n_{HP}	n_P
N	n_{DN}	n_{HN}	n_N
	n_D	n_H	N

Then the estimate of $\Pr(D)$ after N persons is n_D/N, of $\Pr(D|P)$ is n_{DP}/n_P and so on.

Because $\Pr(D)$ ($= 0.02$) is low the probability estimates are not evaluated until $N \geqslant 200$ and $n_D > 0$ (statement 580) to avoid division by zero in evaluating the probability estimates.

Program listings and sample outputs
(a) Program 3.1

```
100 PRINT"[]":REM**CLEAR SCREEN
110 PRINT"       **PROGRAM 3.1**"
120 PRINT" COIN-TOSSING SIMULATION"
130 PRINT:PRINT
140 REM**SIMULATION OF N TOSSES OF COIN WITH A
150 REM**PROBABILITY P OF OBTAINING HEADS.
160 REM**PROBABILITY ESTIMATES PHAT TO BE PRINTED OUT
170 REM**AT INTERVALS OF K TOSSES
180 XO=RND(-2):REM**INITIALISES RANDOM NUMBER SEQUENCE
190 READ N,P K
200 FOR I=1 TO N
210 U=RND(1):REM**GENERATE A RANDOM NUMBER U (0<=U<=1)
220 REM**IF 0<=U<=P THEN HEADS
230 IF U<=P THEN NH=NH+1
240 REM**CALCULATE THE CURRENT ESTIMATE OF THE PROBABILITY OF HEADS
250 PHAT=INT(1000*NH/I+0.5)/1000
260 REM**DETERMINE IF PHAT-P HAS CHANGED SIGN
270 S2=SGN(NH-I*P)
280 IF I=1 THEN 330
290 IF S2=S1 THEN 330
300 IF S2=-1 THEN PRINT"*ON THE";I;"TH TOSS PHAT DROPS BELOW ";P
310 IF S2=0 THEN PRINT"*ON THE";I;"TH TOSS PHAT=";P
320 IF S2=1 THEN PRINT"*ON THE";I;"TH TOSS PHAT INCREASES ABOVE ";P
330 S1=S2
340 IF I=INT(I/K)*K THEN PRINT"ON THE";I;"TH TOSS PHAT=";PHAT
350 NEXT I
360 DATA 1300,0.5.100
370 END
```

Sample of the output from program 3.1

```
THEORETICAL PROB. OF HEADS IS .5

ON THE 100 TH TOSS PHAT= .54
ON THE 200 TH TOSS PHAT= .535
*ON THE 292 TH TOSS PHAT= .5
*ON THE 293 TH TOSS PHAT DROPS BELOW   .5
*ON THE 294 TH TOSS PHAT= .5
*ON THE 295 TH TOSS PHAT DROPS BELOW   .5
ON THE 300 TH TOSS PHAT= .493
ON THE 400 TH TOSS PHAT= .488
ON THE 500 TH TOSS PHAT= .478
ON THE 600 TH TOSS PHAT= .477
ON THE 700 TH TOSS PHAT= .48
ON THE 800 TH TOSS PHAT= .488
ON THE 900 TH TOSS PHAT= .486
ON THE 1000 TH TOSS PHAT= .494
ON THE 1100 TH TOSS PHAT= .495
ON THE 1200 TH TOSS PHAT= .498
*ON THE 1206 TH TOSS PHAT= .5
*ON THE 1207 TH TOSS PHAT INCREASES ABOVE   .5
*ON THE 1208 TH TOSS PHAT= .5
*ON THE 1209 TH TOSS PHAT INCREASES ABOVE   .5
*ON THE 1210 TH TOSS PHAT= .5
*ON THE 1211 TH TOSS PHAT DROPS BELOW   .5
*ON THE 1228 TH TOSS PHAT= .5
*ON THE 1229 TH TOSS PHAT DROPS BELOW   .5
*ON THE 1230 TH TOSS PHAT= .5
*ON THE 1231 TH TOSS PHAT DROPS BELOW   .5
*ON THE 1270 TH TOSS PHAT= .5
*ON THE 1271 TH TOSS PHAT INCREASES ABOVE   .5
*ON THE 1272 TH TOSS PHAT= .5
*ON THE 1273 TH TOSS PHAT DROPS BELOW   .5
ON THE 1300 TH TOSS PHAT= .496
```

(b) Program 3.2

```
100 PRINT"█":REM**CLEAR SCREEN
110 PRINT"           **PROGRAM 3.2**"
120 PRINT"        SWITCHING SIMULATION"
130 PRINT:PRINT
140 REM**SIMULATION OF EXAMPLES 3.2,3.3 AND 3.4
150 REM**N REPETITIONS OF EXPERIMENT
160 REM**ESTIMATES PRINTED AT INTERVALS OF K REPETITIONS
170 X0=RND(-2):REM**INITIALISE RANDOM NUMBER GENERATOR
180 REM**ZERO EVENT COUNTS FOR:-
190 REM**(A)FLOW OF CURRENT FOR SERIES SWITCHING (N1S)
200 REM**(B)FLOW OF CURRENT FOR PARALLEL SWITCHING (N2P)
210 REM**(C)S1 BEING CLOSED WHEN A CURRENT CAN FLOW (PARALLEL CASE)
220 N1S=0 : N2P=0 : N3S1=0
230 REM**READ PI(PROB. SWITCH I(I=1,2)IS CLOSED)N AND K
240 READ P1,P2,N,K
250 REM**CALCULATION OF THEORETICAL PROBABILITIES Q1,Q2,Q3
260 Q1=P1*P2 : Q2=P1+P2-Q1 : Q3=P1/Q2
265 FOR I=1 TO N
270 REM**SET SWITCHES(1 CLOSED,0 OPEN)
280 U1=RND(1) : C1=0 : IF U1<=P1 THEN C1=1
290 U2=RND(1) : C2=0 : IF U2<=P2 THEN C2=1
300 REM**TEST FOR CURRENT FLOW (SERIES SWITCHES)
310 IF C1=1 AND C2=1 THEN N1S=N1S+1
320 REM**TEST FOR CURRENT FLOW (PARALLEL SWITCHES)
330 IF C1=1 OR C2=1 THEN N2P=N2P+1
340 REM**TOTAL REPETITIONS WHERE S1 IS CLOSED
350 IF C1=1 THEN N3S1=N3S1+1
360 REM**CALCULATE PROBABILITY ESTIMATES FOR (A),(B),(C)
370 E1S=INT(1000*N1S/I+0.5)/1000
380 E2P=INT(1000*N2P/I+0.5)/1000
390 E3S1=INT(1000*N3S1/N2P+0.5)/1000
400 IF I<>INT(I/K)*K THEN 490
410 PRINT"PROB. EST. AT";I;"TH REP.(&THEOR. VALUES)"
420 PRINT"(A) PR(CURRENT FLOW,S1 AND S2 IN SERIES)=";E1S;
430 PRINT"(";Q1;")"
440 PRINT"(B) PR(CURRENT FLOW,S1 AND S2 IN PARALLEL)=";E2P;
450 PRINT"(";Q2;")"
460 PRINT"(C) PR(S1 OPEN/CURRENT FLOWS)=";E3S1;
470 PRINT"(";INT(1000*Q3+0.5)/1000;")"
480 PRINT
490 REM
500 NEXT I
510 DATA 0.3,0.4,1000,100
1000 END
```

Sample of the output from program 3.2

```
PROB. S1 CLOSED= .3
PROB. S2 CLOSED= .4


PROB. EST. AT 100 TH REP.(&THEOR. VALUES)
(A) PR(CURRENT FLOW,S1 AND S2 IN SERIES)= .17 ( .12 )
(B) PR(CURRENT FLOW,S1 AND S2 IN PARALLEL)= .62 ( .58 )
(C) PR(S1 OPEN/CURRENT FLOWS)= .532 ( .517 )

PROB. EST. AT 200 TH REP.(&THEOR. VALUES)
(A) PR(CURRENT FLOW,S1 AND S2 IN SERIES)= .12 ( .12 )
(B) PR(CURRENT FLOW,S1 AND S2 IN PARALLEL)= .575 ( .58 )
(C) PR(S1 OPEN/CURRENT FLOWS)= .583 ( .517 )

PROB. EST. AT 300 TH REP.(&THEOR. VALUES)
(A) PR(CURRENT FLOW,S1 AND S2 IN SERIES)= .107 ( .12 )
(B) PR(CURRENT FLOW,S1 AND S2 IN PARALLEL)= .567 ( .58 )
(C) PR(S1 OPEN/CURRENT FLOWS)= .576 ( .517 )

PROB. EST. AT 400 TH REP.(&THEOR. VALUES)
(A) PR(CURRENT FLOW,S1 AND S2 IN SERIES)= .105 ( .12 )
(B) PR(CURRENT FLOW,S1 AND S2 IN PARALLEL)= .57 ( .58 )
(C) PR(S1 OPEN/CURRENT FLOWS)= .553 ( .517 )

PROB. EST. AT 500 TH REP.(&THEOR. VALUES)
(A) PR(CURRENT FLOW,S1 AND S2 IN SERIES)= .104 ( .12 )
(B) PR(CURRENT FLOW,S1 AND S2 IN PARALLEL)= .576 ( .58 )
(C) PR(S1 OPEN/CURRENT FLOWS)= .528 ( .517 )

PROB. EST. AT 600 TH REP.(&THEOR. VALUES)
(A) PR(CURRENT FLOW,S1 AND S2 IN SERIES)= .102 ( .12 )
(B) PR(CURRENT FLOW,S1 AND S2 IN PARALLEL)= .57 ( .58 )
(C) PR(S1 OPEN/CURRENT FLOWS)= .515 ( .517 )

PROB. EST. AT 700 TH REP.(&THEOR. VALUES)
(A) PR(CURRENT FLOW,S1 AND S2 IN SERIES)= .101 ( .12 )
(B) PR(CURRENT FLOW,S1 AND S2 IN PARALLEL)= .579 ( .58 )
(C) PR(S1 OPEN/CURRENT FLOWS)= .511 ( .517 )

PROB. EST. AT 800 TH REP.(&THEOR. VALUES)
(A) PR(CURRENT FLOW,S1 AND S2 IN SERIES)= .103 ( .12 )
(B) PR(CURRENT FLOW,S1 AND S2 IN PARALLEL)= .585 ( .58 )
(C) PR(S1 OPEN/CURRENT FLOWS)= .509 ( .517 )

PROB. EST. AT 900 TH REP.(&THEOR. VALUES)
(A) PR(CURRENT FLOW,S1 AND S2 IN SERIES)= .104 ( .12 )
(B) PR(CURRENT FLOW,S1 AND S2 IN PARALLEL)= .576 ( .58 )
(C) PR(S1 OPEN/CURRENT FLOWS)= .508 ( .517 )

PROB. EST. AT 1000 TH REP.(&THEOR. VALUES)
(A) PR(CURRENT FLOW,S1 AND S2 IN SERIES)= .107 ( .12 )
(B) PR(CURRENT FLOW,S1 AND S2 IN PARALLEL)= .579 ( .58 )
(C) PR(S1 OPEN/CURRENT FLOWS)= .504 ( .517 )
```

(c) Program 3.3

```
100 PRINT"█":REM**CLEAR SCREEN
110 PRINT"        **PROGRAM 3.3**"
120 PRINT"ILLUSTRATION OF BAYES' THEOREM"
130 PRINT"        EXAMPLE 3.6"
140 PRINT:PRINT
150 REM**PROB. ESTS. PRINTED OUT EVERY K PATIENTS
160 REM**D-DISEASED;H-HEALTHY
170 REM**P-POSITIVE TEST;N-NEGATIVE TEST
180 REM**P1=PR(D);P2=PR(P/D);P3=PR(N/H)
190 READ N,K,P1,P2,P3
200 REM**CALCULATION OF THEOR. PROBS.
210 REM**P5=PR(D/P);P6=PR(H/P);P7=PR(D/N);P8=PR(H/N)
220 P5=INT(10000*P2*P1/(P2*P1+(1-P3)*(1-P1))+0.5)/10000
230 P6=1-P5
240 P7=INT(10000*(1-P2)*P1/((1-P2)*P1+P3*(1-P1))+0.5)/10000
250 P8=1-P7
260 X0=RND(-2):REM**INITIALISES RANDOM NUMBER SEQUENCE
270 REM**ZERO COUNTS FOR
280 REM**(1) N1D-NO. OF DISEASED PATIENTS
290 REM**(2) N2H-NO.OF HEALTHY PATIENTS
300 REM**(3) N3P-NO. OF POSITIVE TESTS
310 REM**(4) N4N-NO. OF NEGATIVE TESTS
320 REM**(5) N5DP-NO. OF POS. TESTS ASSOCIATED
330 REM**WITH DISEASED PATIENTS;
340 REM**N6HP,N7DN,N8HN SIMILARLY DEFINED
350 N1D=0 : N2H=0 : N3P=0 : N4N=0
360 N5DP=0 : N6HP=0 : N7DN=0 : N8HN=0
370 FOR I=1 TO N
380 REM**CURRENT PATIENT DISEASED (Q1=1) OR HEALTHY (Q1=0)?
390 U1=RND(1) : Q1=0 : IF U1<=P1 THEN Q1=1
400 REM**TEST RESULT POS. (Q2=1) OR NEG. (Q2=0)?
410 U2=RND(1)
420 REM**DISEASED PATIENT
430 IF Q1=1 AND U2<=P2 THEN Q2=1
440 IF Q1=1 AND U2>P2 THEN Q2=0
450 REM**HEALTHY PATIENT
460 IF Q1=0 AND U2>P3 THEN Q2=1
470 IF Q1=0 AND U2<=P3 THEN Q2=0
480 REM**UPDATE NUMBERS IN DIFFERENT CATEGORIES
490 IF Q1=1 THEN N1D=N1D+1
500 IF Q1=0 THEN N2H=N2H+1
510 IF Q2=1 THEN N3P=N3P+1
520 IF Q2=0 THEN N4N=N4N+1
530 IF Q1=1 AND Q2=1 THEN N5DP=N5DP+1
540 IF Q1=0 AND Q2=1 THEN N6HP=N6HP+1
550 IF Q1=1 AND Q2=0 THEN N7DN=N7DN+1
560 IF Q1=0 AND Q2=0 THEN N8HN=N8HN+1
570 REM**CALCULATION OF PROB. ESTS.
580 IF I<200 OR N1D=0 THEN 780
590 REM**AVOIDS DIVISION BY ZERO
600 E1D=INT(1000*N1D/I+0.5)/1000
610 E2PD=INT(1000*N5DP/N1D+0.5)/1000
620 E3NH=INT(1000*N8HN/N2H+0.5)/1000
630 E5DP=INT(1000*N5DP/N3P+0.5)/1000
640 E6HP=INT(1000*N6HP/N3P+0.5)/1000
650 E7DN=INT(10000*N7DN/N4N+0.5)/10000
660 E8HN=INT(1000*N8HN/N4N+0.5)/1000
670 IF I<>INT(I/K)*K THEN 780
680 PRINT"PROB.ESTS. AFTER";I;"TH PATIENT(+THEOR. VALUES) "
690 PRINT"D-DISEASED:H-HEALTHY:P-POS. TEST:N-NEG. TEST"
700 PRINT"PR(D)=";E1D;"(";P1;")"
710 PRINT"PR(P/D)=";E2PD;"(";P2;")"
720 PRINT"PR(N/H)=";E3NH;"(";P3;")"
```

```
730 PRINT"PR(D/P)=";E5DP;"(";P5;")"
740 PRINT"PR(H/P)=";E6HP;"(";P6;")"
750 PRINT"PR(D/N)=";E7DN;"(";P7;")"
760 PRINT"PR(H/N)=";E8HN;"(";P8;")"
770 PRINT
780 REM
790 NEXT I
800 DATA 600,100,0.02,0.75,0.97
810 END
```

Sample of the output from program 3.3

```
PROB.ESTS. AFTER 200 TH PATIENT(+THEOR. VALUES)
D-DISEASED:H-HEALTHY:P-POS. TEST:N-NEG. TEST
PR(D)= 5E-03 ( .02 )
PR(P/D)= 1 ( .75 )
PR(N/H)= .975 ( .97 )
PR(D/P)= .167 ( .3378 )
PR(H/P)= .833 ( .6622 )
PR(D/N)= 0 ( 5.2E-03 )
PR(H/N)= 1 ( .9948 )

PROB.ESTS. AFTER 300 TH PATIENT(+THEOR. VALUES)
D-DISEASED:H-HEALTHY:P-POS. TEST:N-NEG. TEST
PR(D)= 7E-03 ( .02 )
PR(P/D)= .5 ( .75 )
PR(N/H)= .97 ( .97 )
PR(D/P)= .091 ( .3378 )
PR(H/P)= .909 ( .6622 )
PR(D/N)= 3.5E-03 ( 5.2E-03 )
PR(H/N)= .997 ( .9948 )

PROB.ESTS. AFTER 400 TH PATIENT(+THEOR. VALUES)
D-DISEASED:H-HEALTHY:P-POS. TEST:N-NEG. TEST
PR(D)= .013 ( .02 )
PR(P/D)= .8 ( .75 )
PR(N/H)= .959 ( .97 )
PR(D/P)= .19 ( .3378 )
PR(H/P)= .81 ( .6622 )
PR(D/N)= 2.6E-03 ( 5.2E-03 )
PR(H/N)= .997 ( .9948 )

PROB.ESTS. AFTER 500 TH PATIENT(+THEOR. VALUES)
D-DISEASED:H-HEALTHY:P-POS. TEST:N-NEG. TEST
PR(D)= .012 ( .02 )
PR(P/D)= .833 ( .75 )
PR(N/H)= .966 ( .97 )
PR(D/P)= .217 ( .3378 )
PR(H/P)= .783 ( .6622 )
PR(D/N)= 2.1E-03 ( 5.2E-03 )
PR(H/N)= .998 ( .9948 )

PROB.ESTS. AFTER 600 TH PATIENT(+THEOR. VALUES)
D-DISEASED:H-HEALTHY:P-POS. TEST:N-NEG. TEST
PR(D)= .012 ( .02 )
PR(P/D)= .857 ( .75 )
PR(N/H)= .963 ( .97 )
PR(D/P)= .207 ( .3378 )
PR(H/P)= .793 ( .6622 )
PR(D/N)= 1.8E-03 ( 5.2E-03 )
PR(H/N)= .998 ( .9948 )
```

APPENDIX 3.1 PROOF OF BAYES' THEOREM ($k = 2$)

When $k = 2$, E_1 and E_2 form the mutually exclusive exhaustive set. From the frequency definition of probability

$$\Pr(E_i|A) = \frac{n_{AE_i}(L)}{n_A(L)}$$

(where out of N repetitions, the event AE_i occurs n_{AE_i} times, and the event A occurs n_A times)

$$= \frac{n_{AE_i}(L)/N(L)}{n_A(L)/N(L)} \qquad \left(= \frac{\Pr(AE_i)}{\Pr(A)}\right)$$

dividing numerator and denominator by $N(L)$.

Since E_1 and E_2, and hence AE_1 and AE_2, are mutually exclusive, $n_A = n_{AE_1} + n_{AE_2}$. Hence

$$\Pr(E_i|A) = \frac{n_{AE_i}(L)/N(L)}{(n_{AE_1}(L) + n_{AE_2}(L))/N(L)} \left(= \frac{\Pr(AE_i)}{\Pr(AE_1) + \Pr(AE_2)}\right)$$

Now we can write $n_{AE_i}(L)/N(L) = n_{AE_i}(L)/n_{E_i}(L) \times n_{E_i}(L)/N(L)$; then

$$\Pr(E_i|A) = \frac{n_{AE_i}(L)/n_{E_i}(L) \times n_{E_i}(L)/N(L)}{n_{AE_1}(L)/n_{E_1}(L) \times n_{E_1}(L)/N(L) + n_{AE_2}(L)/n_{E_2}(L) \times n_{E_2}(L)/N(L)}$$

$$= \frac{\Pr(A|E_i) \times \Pr(E_i)}{\Pr(A|E_1) \times \Pr(E_1) + \Pr(A|E_2) \times \Pr(E_2)} \qquad i = 1, 2 . \qquad (3.18)$$

EXERCISES

Section A

1. The weights (in grams) of a hundred sodium bicarbonate tablets may be reduced to the distribution below:

Class mid-value	0.3295	0.3495	0.3695	0.3895	0.4095	0.4295	0.4495	0.4695	0.4895
Frequency	1	0	8	10	22	23	29	6	1

Let E_1, E_2 and E_3 be the events that the weight of a tablet lies in the intervals 0.3595 to 0.4395 g, 0.4595 to 0.4795 g and 0.3995 to 0.4595 g respectively. Estimate the probabilities (a) $\Pr(E_1)$, (b) $\Pr(E_2)$, (c) $\Pr(E_3)$, (d) $\Pr(E_1 E_2)$, (e) $\Pr(E_1 + E_2)$, (f) $\Pr(E_1 E_3)$, (g) $\Pr(E_1 + E_3)$.

2. Two events A and B are mutually exclusive; $\Pr(A) = 1/5$ and $\Pr(B) = 1/3$. Find the probability that (a) either A or B will occur, (b) both A and B will occur, (c) neither A nor B will occur.

3. Two events A and B are independent; $\Pr(A) = 0.3$ and $\Pr(B) = 0.5$. Find (a) $\Pr(AB)$, (b) the probability that neither A nor B will occur, (c) $\Pr(A + B)$.

4. A bottle of tablets contains a proportion 0.1 of white tablets, 0.2 of blue, 0.3 of red and 0.4 of yellow. The white and yellow tablets contain codeine while the others do not. A tablet is drawn at random from the bottle. Let E_1 be the event that the tablet is white or blue and E_2 be the event that the tablet contains codeine. Find (a) $\Pr(E_1)$, (b) $\Pr(E_2)$, (c) $\Pr(E_1 E_2)$, (d) $\Pr(E_1 | E_2)$, (e) $\Pr(E_1 + E_2)$.

5. The probability that a sample of a mixture A contains more than a critical level of impurity is 0.1, and for a sample of mixture B the probability is 0.2. If one sample of each mixture is chosen at random find the probabilities that (a) neither, (b) one of them, (c) both, contain more than the critical level of the impurity.

6. Out of 12 samples of a mixture, four are known to contain more than a certain critical level of impurity. If two of the samples are chosen at random (and without replacement) find the probabilities that (a) the first one only contains more than the critical level, (b) exactly one contains more than the critical level, (c) at least one contains more than the critical level. Repeat the calculations if three of the samples are chosen at random.

7. Immigration from four countries A, B, C, D occurs in the ratio 7:4:3:2. Two immigrants from the same country are tested for the presence of a certain disease, which is detected in one of them, the other being cleared. Unfortunately the records stating their country of origin were mislaid. Find the probability that they come from country D assuming that the probability of a person's contracting the disease in A, B, C, D is 0.1, 0.2, 0.25 and 0.4 respectively. Find also the country which they are most likely to have come from.

Section B

1. Out of ten cars three are known to have a certain fault. If two of the cars are chosen at random find the probability that (a) the first car is faulty and the second car is not, (b) one of the cars is faulty, (c) both cars are faulty.

2. Thirty per cent of a large batch of cars are known to possess a certain fault. Find the probability that if three of the cars are selected at random they will all be faulty.

3. A's chance of completing an experiment within two hours is 5/7 and B's and C's chances are 4/7 and 3/7 respectively. What are the chances of at least two of them completing the experiment within the allotted time?

4. For a system to work satisfactorily two independent components A and B must both work satisfactorily. If the probability that A fails is 0.01 and that

B fails is 0.02 find the probability that the system works satisfactorily. Find the probability that component A has failed given that the system has failed.

5. The switches in the circuit shown in the diagram are independent. For each switch the probability of being closed is 0.1. Construct a tree diagram to illustrate the elementary events associated with the system. Calculate, using two different approaches, the probability that a current can flow from A to B.

Find the probability that switch S_1 is closed given that a current can flow from A to B.

.6. A certain transistor is manufactured at three different factories at Barnsley, Bradford and Bristol. It is known that the Barnsley factory produces twice as many transistors as the Bradford one, which produces the same number as the Bristol one (during the same period). Experience also shows that 0.2% of the transistors produced at Barnsley and Bradford are faulty and so are 0.4% of those produced at Bristol.

The outputs of all three factories are mixed up in a single store at a factory in Birmingham which produces electronic instruments each of which contains one transistor.

A purchaser finds that his instrument is faulty, and the service engineer traces this fault to a defective transistor. What is the probability that the Bradford factory is to blame?

Section C

1. Run program 3.1 using different random number sequences. If $p = 0.5$ estimate the number of throws necessary to get within (a) 0.1, (b) 0.01 of p. Also examine the variation of the interval between the times at which the estimate of the probability coincides with the true value.
2. Write a program to simulate the situation of example 3.1.
3. Write a program to simulate the throwing of a given score on a fair six-sided die. Modify the program to allow for the die not being fair. If the probability of getting a six on the die is 1/7 instead of 1/6, estimate from runs of the program how many throws of the die will be necessary before this difference is detected.
4. Write a computer program to simulate the situation of example 3.5. Hence verify the values for the probabilities obtained in the solution to this example.

4

Probability Distributions

4.1 INTRODUCTION

In the introduction to Chapter 3 we discussed several examples of random experiments. The possible outcomes of these experiments can be described in terms of random variables as follows.

Experiment 1. Tossing a coin. The outcomes of the experiment can be described in terms of X, the number of heads obtained in a toss. The possible values of X are $X = 0$ (a tail) and $X = 1$ (a head).

Experiment 2. Tossing three coins simultaneously. Here the random variable can be taken as X, the number of heads obtained, with possible values $X = 0, 1, 2$ and 3.

Experiment 3. Counting the number of disintegrations from a radioactive source in a given interval of time. This number X is a random variable with possible values $X = 0, 1, 2, 3, \ldots$; in theory at least there is no upper limit to the value which X can take.

Experiment 4. Measuring the yield Y g of a product in a chemical reaction. The random variable Y lies in the range $Y \geqslant 0$.

Experiment 5. Weighing four aspirin tablets and finding their mean weight $\bar{X}$ g. The random variable $\bar{X}$ lies in the range $\bar{X} \geqslant 0$.

In none of the above experiments can we predict with certainty the value which the random variable will take on any repetition of the experiment. However, what can be done is to specify the probability that the random variable takes a particular value, or lies in a particular range, before the experiment has been performed.

The outcomes of experiments 1, 2 and 3 are described in terms of *discrete random variables* whose ranges consist of a set of discrete values. The *probability distributions* of discrete random variables are specified in terms of *probability mass functions*.

The outcomes of experiments 4 and 5 are described in terms of *continuous random variables* which can take values on some continuous interval. The probability distribution of a continuous random variable is specified in terms of a *probability density function.*

In sections 4.2, 4.3 and 4.4 the distributions of discrete random variables which have *binomial, geometric* and *Poisson* distributions will be discussed. Sections 4.5 and 4.6 are devoted to the discussion of the *negative exponential* and *normal* distributions. In section 4.7 the Poisson approximation to the binomial distribution is discussed, and the normal approximation to both the binomial and Poisson distributions is introduced. The chapter concludes with a section on the generation of data and probabilities according to the binomial, Poisson and normal distributions, using desk-top computers.

4.2 THE BINOMIAL DISTRIBUTION

4.2.1 Probability mass function for the binomial distribution

Consider the following problem:

Twenty-five per cent of the trout in a lake are known to be infested by a particular type of parasite. If a random sample of 3 trout are caught from the lake find the probability distribution of the number of infested trout in the sample.

Here the random variable X is the number of infested trout in the sample. It is a discrete random variable which can take on the values, 0, 1, 2 and 3. The set of probabilities

$$\Pr(X = 0), \quad \Pr(X = 1), \quad \Pr(X = 2) \quad \text{and} \quad \Pr(X = 3)$$

defines the *probability mass function* of X.

Catching a fish is a 'trial' so that in obtaining a sample of 3 fish, 3 trials are involved. A trial here is assumed to be the equivalent of drawing a trout at random from the trout population in the lake; that is at each trial each trout has an equal chance of being caught. It is also assumed that whether or not a trout is infested does not affect the probability of it being caught. Also we assume that the probability of catching an infested trout remains constant (= 1/4) at each trial. This is equivalent to the assumption that the sample size is very much smaller than the trout population in the lake, so that the removal of the sample, or part of it, has a negligible effect on the distribution of the infestation in the remaining fish. It is further assumed that the results of the trials are independent.

Under the conditions discussed above it is possible to evaluate the probability mass function of X. As an illustration we shall calculate $\Pr(X = 2)$ and then write down the generalisation of the result obtained.

Let I_j be the event that an infested fish is caught on the jth trial ($j = 1, 2, 3$) and N_j the event that a non-infested fish is caught on the trial. Then

$$
\begin{aligned}
\Pr(X = 2) &= \Pr(I_1 I_2 N_3 + I_1 N_2 I_3 + N_1 I_2 I_3) \\
&= \Pr(I_1 I_2 N_3) + \Pr(I_1 N_2 I_3) + \Pr(N_1 I_2 I_3) \text{ since the} \\
&\qquad\qquad\qquad\qquad\qquad\qquad \text{events are mutually exclusive} \\
&= \Pr(I_1) \times \Pr(I_2) \times \Pr(N_3) + \Pr(I_1) \times \Pr(N_2) \times \Pr(I_3) \\
&\quad + \Pr(N_1) \times \Pr(I_2) \times \Pr(I_3)
\end{aligned}
$$

since the trials are independent

$$
= \binom{3}{2} \times \frac{3}{4} \times \left(\frac{1}{4}\right)^2
$$

since the probabilities of catching an infested fish ($= 1/4$) and a non-infested fish ($= 3/4$) remain constant from one trial to the next.

Here $\binom{3}{2}$ (read as '3C2') is the number of ways of choosing the two trials on which infested fish are caught, out of the three trials in the experiment. In general $\binom{n}{r}$ (read as 'nCr') may be evaluated as follows

$$
\binom{n}{r} = \frac{n!}{r!\,(n-r)!} \tag{4.1}
$$

where $n!$ (read as 'factorial n') may be evaluated from

$$
n! = n \times (n-1) \times \ldots \times 3 \times 2 \times 1 \tag{4.2}
$$

$\binom{n}{r}$ is interpreted as the number of different ways of choosing r ($\leqslant n$) places out of n places, and thus both r and n must be integers.

As there is only one way in which either 0 or n places may be chosen from n places we define

$$
\binom{n}{0} = \binom{n}{n} = 1 \tag{4.3}
$$

From (4.1) $\binom{3}{2} = \dfrac{3!}{2!\,1!} = 3$, which is of course in agreement with the result already derived from first principles.

The result obtained for $\Pr(X = 2)$ may be generalised as follows. Consider an experiment which consists of n independent trials. Each trial has two possible outcomes, 'success' (with probability p) and 'failure' (with probability $q = 1 - p$).

Trials of this type are referred to as *Bernoulli trials*. Let X be the number of successes in the experiment. Then X is a random variable which can take on one of the values $0, 1, 2, \ldots, n$. It can then be shown by arguments similar to those used in the particular example above that

$$\Pr(X = r) = \binom{n}{r} q^{n-r} p^r \qquad r = 0, 1, 2, \ldots, n \qquad (4.4)$$

The random variable X is said to have a *binomial distribution;* this nomenclature arises from the fact that the probabilities given by (4.4) also occur as the terms in the binomial expansion of $(q + p)^n$. The values of the binomial probabilities are completely specified in terms of n (the number of trials in the experiment) and p (the probability of the outcome labelled 'success' at each trial). Thus we write

$$X \sim b(n, p)$$

which reads 'X has a binomial distribution with parameters n and p'.

Ex. 4.1. Evaluate (i) $\binom{6}{4}$, (ii) $\binom{100}{99}$.

(i) $\binom{6}{4} = \dfrac{6!}{4!2!} = \dfrac{6 \times 5 \times 4!}{4! \times 2!} = \dfrac{6 \times 5}{2 \times 1} = 15$

(ii) $\binom{100}{99} = \dfrac{100!}{99!1!} = \dfrac{100 \times 99!}{99! \times 1!} = 100$

Note that in the above use has been made of the fact that the factorial of the larger number in the denominator will divide into the factorial in the numerator.

Ex. 4.2. Using (4.4) find the probability distribution of the random variable X in the introductory problem to section 4.2.1.

Here n is the sample size, 3, and p is the probability of catching an infested fish, that is $1/4$. Therefore $X \sim b(3, 1/4)$ and

$$\Pr(X = r) = \binom{3}{r}\left(\frac{3}{4}\right)^{3-r}\left(\frac{1}{4}\right)^r \qquad r = 0, 1, 2, 3 \ .$$

$$\Pr(X = 0) = \binom{3}{0}\left(\frac{3}{4}\right)^3\left(\frac{1}{4}\right)^0 = \frac{27}{64} \ ,$$

using $\binom{3}{0} = 1$ from (4.3), and $a^0 = 1$ (for all $a > 0$) .

$$\Pr(X = 1) = \binom{3}{1}\left(\frac{3}{4}\right)^2\left(\frac{1}{4}\right)^1 = \frac{27}{64}$$

$$\Pr(X = 2) = \binom{3}{2}\left(\frac{3}{4}\right)^1\left(\frac{1}{4}\right)^2 = \frac{9}{64}$$

$$\Pr(X = 3) = \binom{3}{3}\left(\frac{3}{4}\right)^0\left(\frac{1}{4}\right)^3 = \frac{1}{64}$$

Note that in general, since $\binom{n}{0} = \binom{n}{n} = 1$ and $q^0 = p^0 = 1$,

$$\Pr(X = 0) = q^n \text{ and } \Pr(X = n) = p^n.$$

Examples of random variables whose probability distributions are binomial, are

(i)   the number of heads in n tosses of a coin
(ii)   the number of sixes in n tosses of a die
(iii)   the number of members in a sample with a given characteristic when the sample is drawn at random from a large population, a fixed proportion of whose members have that characteristic.

(The last example has important applications in the quality control of industrial processes and in the analysis of the results of surveys.)

(iv)   the number of successes in an experiment consisting of a fixed number of trials, where the probability of success is constant from trial to trial and the results of the trials are independent.

Consider an experiment whose outcome can be described in terms of the random variable $X \sim b(n, p)$. If the experiment is repeated N times, the results can be expressed in terms of the frequency distribution

X	0	1	2	...	n	
f	f_0	f_1	f_2	...	f_n	N

As N becomes very large the ratio f_r/N will tend towards $\Pr(X = r)$, and the mean and variance of the frequency distribution will tend towards the mean μ and variance σ^2 of the random variable X. It can be shown from theoretical considerations (see Appendix 4.2) that the mean and variance for the binomial distribution are given by

$$\mu = np \tag{4.5}$$

$$\sigma^2 = np(1 - p) = npq \tag{4.6}$$

The mean is also referred to as the *expected value* of X (written as $E(X)$).

Ex. 4.3. In the manufacture of containers by a certain process it is known that 5 per cent of the containers are defective because they fail to satisfy tolerance requirements. What is the probability that a sample of 12 containers selected at random from a large batch contains (a) exactly 2, (b) not more than 2, (c) at least 2 defectives? Find the mean and variance of the number of defectives in the sample.

If the sample size (12) is small compared with the batch size (in practice less than one-tenth of the batch size), it can be assumed that the probability of drawing a defective container remains constant $(= 0.05)$ during the process of sampling the 12 containers, and that the trials are independent. Under these conditions

$$X \sim b(12, 0.05)$$

where X is the number of defective items in the sample. Then

$$\Pr(X = r) = \binom{12}{r} 0.95^{12-r} 0.05^r \quad r = 0, 1, \ldots, 12 \; .$$

(a) $\Pr(X = 2) \; = \; \binom{12}{2} 0.95^{10} \, 0.05^2 = 0.099 \; .$

(b) $\Pr(X \leqslant 2) \; = \; \Pr(X = 0) + \Pr(X = 1) + \Pr(X = 2)$

$$= \; 0.95^{12} + \binom{12}{1} 0.95^{11} \, 0.05 + \binom{12}{2} 0.95^{10} \, 0.05^2$$

$$= \; 0.540 + 0.341 + 0.099$$

$$= \; 0.98 \; .$$

(c) $\Pr(X \geqslant 2) \; = \; 1 - \Pr(X = 0) - \Pr(X = 1), \text{ since } \sum_{r=0}^{12} \Pr(X = r) = 1$

$$= \; 1 - 0.341 - 0.540$$

$$= \; 0.12 \; .$$

From (4.5) the mean, or expected value, of X is given by

$$E(X) = 12 \times 0.05 = 0.60$$

and from (4.6) the variance of X is

$$\sigma^2 = 12 \times 0.05 \times 0.95 = 0.57.$$

4.2.2 Fitting a binomial distribution to a given set of data

The method of fitting a binomial distribution to a given frequency distribution is illustrated in the example below.

Ex. 4.4. Two hundred samples, each of five fish from a lake, are tested to see if the mercury levels in the fish exceeded a particular value. The frequency distribution of the number X of fish in the sample whose mercury levels exceeded this value is tabulated below

X	0	1	2	3	4	5
f	4	43	61	56	30	6

Fit a binomial distribution to the data.

Suppose $X \sim b(n, p)$. Then we know that $n = 5$, but before we can fit the binomial distribution we have to estimate p from the data. We know that for a binomial distribution $\mu = np$, so that it is reasonable to estimate p from $\bar{x} = n\hat{p}$, where $\bar{x}$ is the mean of the frequency distribution

x	f	fx
0	4	0
1	43	43
2	61	122
3	56	168
4	30	120
5	6	30
	200	483

Then $2.415 = 5\hat{p}$

and $\hat{p} = 0.483$
$\hat{q} = 1 - \hat{p} = 0.517$

Σfx

$2.415 \quad \bar{x}$

For the binomial distribution $\Pr(X = r) = \binom{n}{r} q^{n-r} p^r$. Then

$$\Pr(X = r + 1) = \binom{n}{r+1} q^{n-r-1} p^{r+1}$$

Hence
$$\frac{\Pr(X = r + 1)}{\Pr(X = r)} = \frac{\binom{n}{r+1} q^{n-r-1} p^{r+1}}{\binom{n}{r} q^{n-r} p^r}$$

$$= \frac{n!}{(r+1)!(n-r-1)!} \frac{r!(n-r)!}{n!} \frac{p}{q} = \frac{n-r}{r+1} \frac{p}{q}$$

or

$$Pr(X = r + 1) = \frac{n - r}{r + 1} \frac{p}{q} \; Pr(X = r) \; . \tag{4.7}$$

To work out binomial probabilities it is simpler to evaluate $Pr(X = 0) = q^n$ and to generate the remaining probabilities iteratively using the above result, rather than to generate the probabilities individually using formula (4.4).

For the data of this problem we have

$$Pr(X = r + 1) = \frac{5 - r}{r + 1} \times \frac{0.483}{0.517} \times Pr(X = r)$$

$$= \frac{5 - r}{r + 1} \times 0.93424 \times Pr(X = r)$$

$$
\begin{aligned}
Pr(X = 0) &= (0.517)^5 &&= 0.036936 \\
Pr(X = 1) &= 5 \times 0.93424 \times Pr(X = 0) &&= 0.17254 \\
Pr(X = 2) &= \frac{4}{2} \times 0.93424 \times Pr(X = 1) &&= 0.32238 \\
Pr(X = 3) &= \frac{3}{3} \times 0.93424 \times Pr(X = 2) &&= 0.30118 \\
Pr(X = 4) &= \frac{2}{4} \times 0.93424 \times Pr(X = 3) &&= 0.14069 \\
Pr(X = 5) &= \frac{1}{5} \times 0.93424 \times Pr(X = 4) &&= 0.02629 \; .
\end{aligned}
$$

As a check the sum of the calculated probabilities should be equal to 1. The expected frequency in a class is found by multiplying the above probabilities by the total number of samples (200) leading to

X	e	f
0	7.4	4
1	34.5	43
2	64.5	61
3	60.2	56
4	28.1	30
5	5.3	6

Here the *es* are the expected frequencies in the classes on the assumption of a binomial distribution, and the *fs* are the corresponding observed frequencies. There seems to be reasonable agreement between the two. We shall consider later a statistical test which will give us a criterion to establish whether the fit is satisfactory (section 5.9).

4.3 THE GEOMETRIC DISTRIBUTION

The *geometric distribution* is used to describe the distribution of the number of trials to the first success in a series of Bernoulli trials. Its derivation is illustrated in the example below.

Ex. 4.5. Twenty-five per cent of the fish in a lake are known to be infested by a particular type of parasite. Find the probability distribution for the number of catches until the first infested fish is caught.

Let Y denote the number of catches until the first infested fish is caught. Then the possible values of Y are $1, 2, 3, \ldots$

Let E_r denote the event that an infested fish is caught on the rth trial.

Let $\bar{E}_r$ denote the event that a non-infested fish is caught on the rth trial.

Then

$$
\begin{aligned}
\Pr(Y = 1) \;&= \; \Pr(E_1) \\
&= \; \frac{1}{4} \\
\Pr(Y = 2) \;&= \; \Pr(\bar{E}_1 E_2) \\
&= \; \Pr(\bar{E}_1)\Pr(E_2) \qquad \text{since the trials are independent} \\
&= \; \frac{3}{4} \times \frac{1}{4} \\
\Pr(Y = 3) \;&= \; \Pr(\bar{E}_1 \bar{E}_2 E_3) \\
&= \; \Pr(\bar{E}_1)\;\Pr(\bar{E}_2)\;\Pr(E_3) \\
&= \; \frac{3}{4} \times \frac{3}{4} \times \frac{1}{4} \\
&= \; \left(\frac{3}{4}\right)^2 \frac{1}{4} \; .
\end{aligned}
$$

Continuing in this way we find that

$$
\Pr(Y = r) \; = \; \left(\frac{3}{4}\right)^{r-1} \frac{1}{4} \qquad r = 1, 2, 3, \ldots
$$

The above result may be generalised as follows. In a series of Bernoulli trials with a probability of 'success' p (and of 'failure' $q = 1 - p$) at each trial the distribution of the number of trials Y until the first success is given by the geometric distribution

$$\Pr(Y = r) = q^{r-1}p \qquad r = 1, 2, \ldots \qquad (4.8)$$

It can be shown that the mean of this distribution is $1/p$ and its variance is q/p^2.

A generalisation of the geometric distribution is the *negative binomial distribution* which gives the distribution of the number of trials to the kth ($k = 1, 2, 3, \ldots$) success in a series of Bernoulli trials (see Suggestions for Further Reading, Chapter 4). Thus the geometric distribution is a particular case of the negative binomial distribution with $k = 1$.

Ex. 4.6. Twenty per cent of the items in a large batch are defective. Find the probability that a defective item is first sampled on the fifth draw given that the first two items drawn are non-defective. Compare with the conditional probability that the first defective item is sampled on the third draw.

Let Y be the draw on which the first defective item is sampled. We need

$$\Pr(Y = 5 | Y > 2) = \frac{\Pr(Y = 5, Y > 2)}{\Pr(Y > 2)} \qquad \text{from (3.8)}$$

$$= \frac{\Pr(Y = 5)}{\Pr(Y > 2)}$$

since $Y = 5$ belongs to the set of events $Y > 2$. But

$$\Pr(Y = r) = 0.8^{r-1} (0.2) .$$

Thus

$$\Pr(Y > 2) = 1 - \Pr(Y = 1) - \Pr(Y = 2) = 1 - 0.2 - 0.8 \times 0.2 = 0.64 = 0.8^2$$

and

$$\Pr(Y = 5) = 0.8^4 \times 0.2 ,$$

therefore

$$\Pr(Y = 5 | Y > 2) = \frac{0.8^4 \times 0.2}{0.8^2} = 0.8^2 \times 0.2$$

$$= \Pr(Y = 3) .$$

The above result may be readily extended to show that in a series of Bernoulli trials the conditional probability of obtaining the first success on the $r + s$th trial given that the first s trials are failures is the same as the unconditional probability of obtaining the first success on the rth trial.

4.4 THE POISSON DISTRIBUTION

4.4.1 Probability mass function for the Poisson distribution

Suppose a discrete random variable X has a range of possible values $0, 1, 2, \ldots$. If X has a *Poisson distribution* $(X \sim P(\lambda))$ then

$$Pr(X = r) = e^{-\lambda} \frac{\lambda^r}{r!} \qquad r = 0, 1, 2, \ldots \qquad (4.9)$$

where λ is a positive constant and the number e is 2.7183 (correct to 4 decimal places). Powers of e may be obtained directly from electronic calculators.

In theory there is no upper limit to the value of r but in practice the probabilities become rapidly smaller as r increases above a certain value.

For the Poisson distribution it can be shown (see Appendix 4.3) that $\mu = E(X) = \lambda$ and $\sigma^2 = \lambda$; that is for a Poisson distribution the mean and variance are equal.

Some important applications of the Poisson distribution are to

(1) spatial distribution of plants etc.
(2) radioactive decay
(3) accident occurrences
(4) random arrivals of customers at a service station
(5) machine breakdown.

In (4), for example, we have that under certain conditions (one of which is that the distribution of arrivals in any interval is independent of what has happened in previous intervals), the number of arrivals X in a fixed time interval t is a random variable which has a Poisson distribution. Suppose that the mean rate of arrival of the customers is α/unit time. Then $\mu = \alpha t$ in the time interval t and

$$Pr(X = r) = e^{-\alpha t} \frac{(\alpha t)^r}{r!} \qquad r = 0, 1, 2, \ldots \qquad (4.10)$$

Alternatively the Poisson distribution may be derived as a limiting case of the binomial distribution with $n \to \infty$ and $p \to 0$ in such a way that $np = $ constant; then it can be shown (see Appendix 4.5) that

$$Pr(X = r) = \binom{n}{r} q^{n-r} p^r \to e^{-np} \frac{(np)^r}{r!} \qquad (4.11)$$

In practice if $n \geqslant 50$ and $p \leqslant 0.1$ in such a way that $np \leqslant 5$, we may approximate the distribution $b(n, p)$ by a Poisson distribution with the same mean $(\lambda = np)$.

Ex. 4.7. The number of plants of a certain species at a particular site is known to have a Poisson distribution with a mean of 2 plants/m^2. Find the probability of (a) exactly 2, (b) more than 2, (c) less than 2 plants/m^2.

Let X be the number of plants/m^2. Then

$$\Pr(X = r) = e^{-2} \frac{2^r}{r!}$$

(a) $\Pr(X = 2) = e^{-2} \dfrac{2^2}{2!} = 2e^{-2} = 0.2707$

(b) $\Pr(X > 2) = 1 - \Pr(X = 0) - \Pr(X = 1) - \Pr(X = 2)$ since the probabilities

sum to 1

$$= 1 - e^{-2} (1 + 2 + 2)$$

$$= 0.3233 .$$

(c) $\Pr(X < 2) = \Pr(X = 0) + \Pr(X = 1)$

$$= 3e^{-2}$$

$$= 0.4060 .$$

Ex. 4.8. A sample of flesh from a fish contains strontium 90 and gives a mean count rate of 30 per minute per gram wet weight. If 1/10 g of material is taken what is the probability (a) of more than 3 counts per minute, (b) of obtaining exactly 3 counts in each of two consecutive minutes, (c) of obtaining exactly 6 counts in two minutes?

Let X be the number of counts per minute from 0.1 g of flesh.

Then $E(X) = 30 \times 0.1 = 3$ and $\Pr(X = r) = e^{-3} \dfrac{3^r}{r!}$.

(a) $\Pr(X > 3) = 1 - (\Pr(X = 0) + \Pr(X = 1) + \Pr(X = 2) + \Pr(X = 3))$

$$= 1 - e^{-3} \left(1 + 3 + \frac{3^2}{2!} + \frac{3^3}{3!} \right)$$

$$= 1 - 0.6472$$

$$= 0.3528 .$$

(b) Probability = Pr(3 in first min. and 3 in second min.)

  = Pr(3 in first min.) × Pr(3 in second min.)   since these events
                                                   may be taken to be independent

$$= \left(e^{-3} \, \frac{3^3}{3!} \right)^2$$

$$= (0.2240)^2$$

$$= 0.0502 \; .$$

(c) Let Y be the number of counts in a two-minute period from 0.1 g flesh.

Then $E(Y) = 30 \times 0.1 \times 2 = 6$ and $Pr(Y = r) = e^{-6} \, \dfrac{6^r}{r!}$.

$$Pr(Y = 6) = e^{-6} \, \frac{6^6}{6!}$$

$$= 64.8 e^{-6}$$

$$= 0.1606 \; .$$

4.4.2 Fitting a Poisson distribution to a given set of data

The method of fitting a Poisson distribution to a given frequency distribution is illustrated in the example below. This example also illustrates the use of the Poisson distribution as an approximation to the binomial distribution.

Ex. 4.9. The numbers of defectives in one hundred samples, each of sixty plastic beakers, are noted during a period in which the manufacturing process is under control in order to set up a control chart. The distribution of defectives amongst the samples is tabulated below

X	0	1	2	3	4	5	6
f	11	32	26	14	12	4	1

Fit a Poisson distribution to the data.

Suppose that the proportion of defectives is p; then we would expect $X \sim b(60, p)$. We can estimate p as in example 4.4.

$$\bar{x} = \frac{1}{100} (11 \times 0 + 32 \times 1 + 26 \times 2 + 14 \times 3 + 12 \times 4 + 4 \times 5 + 1 \times 6) = 2$$

Therefore

$$60\hat{p} = 2 \quad \text{or} \quad \hat{p} = 1/30$$

Hence $n > 50, p < 0.1$ with $np < 5$ and the conditions for the Poisson approximation to the binomial distribution are valid. Hence we fit

$$X \sim P(2)$$

If $X \sim P(\lambda)$ we have $\Pr(X = r) = e^{-\lambda} \dfrac{\lambda^r}{r!}$. Hence

$$\frac{\Pr(X = r + 1)}{\Pr(X = r)} = \frac{e^{-\lambda}\lambda^{r+1}}{(r+1)!} \frac{r!}{e^{-\lambda}\lambda^r} = \frac{\lambda}{r+1} \ .$$

Thus

$$\Pr(X = r + 1) = \frac{\lambda}{r+1} \ \Pr(X = r) \ . \tag{4.12}$$

Poisson probabilities are most simply evaluated using the above iterative formula starting with $\Pr(X = 0) = e^{-\lambda}$.

For the data of the example we have

$$\Pr(X = 0) = e^{-2} = 0.1353$$

$$\Pr(X = 1) = \frac{2}{1} \times \Pr(X = 0) = 0.2707$$

$$\Pr(X = 2) = \frac{2}{2} \times \Pr(X = 1) = 0.2707$$

$$\Pr(X = 3) = \frac{2}{3} \times \Pr(X = 2) = 0.1804$$

$$\Pr(X = 4) = \frac{2}{4} \times \Pr(X = 3) = 0.0902$$

$$\Pr(X = 5) = \frac{2}{5} \times \Pr(X = 4) = 0.0361$$

$$\Pr(X = 6) = \frac{2}{6} \times \Pr(X = 5) = 0.0120 \ .$$

Note that $\Pr(X \geqslant 6) = 1 - \Pr(X \leqslant 5) = 0.0166$.

The expected frequencies (e) may be found by multiplying the total number of samples (100) by the above probabilities. Thus we obtain

X	0	1	2	3	4	5	$\geqslant 6$
e	13.5	27.1	27.1	18.0	9.0	3.6	1.7
f	11	32	26	14	12	4	1

Comparison of the observed and expected frequencies shows that the agreement is satisfactory. Later we shall develop a statistical test which will give us a criterion to test the goodness of fit (see section 5.9).

4.5 THE NEGATIVE EXPONENTIAL DISTRIBUTION

4.5.1 Probability density function

Let X be a continuous random variable. Then the statistical properties of X may be determined by specifying the *probability density function* (p.d.f.) of X, $f(x)$ say. This function is defined such that if x to $x + \delta x$ is a small range of possible values of X then

$$\Pr(x < X < x + \delta x) = f(x)\delta x.$$

Thus $f(x)$ is a non-negative function of X.

If we wish to find $\Pr(a < X < b)$ we note that this will be given by the area under the curve $f(x)$ bounded by the ordinates $x = a$ and $x = b$ (Fig. 4.1).

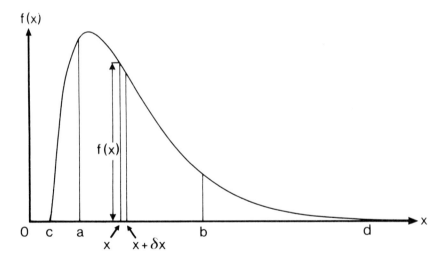

Fig. 4.1 – Probability density function for a continuous random variable X.

Now suppose that c and d are the extreme values of the allowed range for X (where either or both c and d may be infinite); then the area bounded by the curve, the x-axis and the ordinates $x = c$ and $x = d$ will be unity, this follows since X must be somewhere in the range $c < X < d$ so that

$$\Pr(c < X < d) = 1 .$$

The mean and variance σ^2 of X may also be determined using the probability density function $f(x)$. Hence, provided we can specify the p.d.f. of a continuous random variable, we can

(a) calculate the probability that the random variable falls within a given range
(b) calculate the mean of the random variable
(c) calculate the variance of the random variable.

4.5.2 Probability density function for the negative exponential distribution

If a random variable X has a *negative exponential distribution* its probability density function is

$$
\begin{aligned}
f(x) \quad &= \quad \mu e^{-\mu x} \qquad 0 \leqslant x < \infty \\
&= \quad 0 \qquad\qquad \text{otherwise}
\end{aligned}
\qquad (4.13)
$$

where $\mu\ (>0)$ is a constant. Some elementary integration (Appendix 4.4) shows that

(i) $\Pr(0 < X < \infty) = 1$

(ii)   the mean of X is $\dfrac{1}{\mu}$

(iii)   the variance of X is $\dfrac{1}{\mu^2}$

Consider the situation in which the mean number of random events per unit time is α, and the probability of an event occurring in the small interval t to $t + \delta t$ is $\alpha \delta t$ independent of what has happened before time t. Then under these conditions it can be shown that Y, the number of events in time αt, has a Poisson distribution with mean αt $\left(\text{and from (4.10) } \Pr(Y = r) = e^{-\alpha t}\ \dfrac{(\alpha t)^r}{r!} \right)$. Let the random variable T denote the interval between events. Then

$$
\begin{aligned}
\Pr(t < T < t + \delta t) = \quad &\Pr\left((Y = 0 \text{ in the interval } 0 \leqslant T \leqslant t) \text{ and (an event} \right. \\
&\left. \text{occurs in the interval } t \leqslant T \leqslant t + \delta t) \right) \\
= \quad &\Pr(Y = 0 \text{ in the interval } 0 \leqslant T \leqslant t) \\
&\times \Pr(\text{an event occurs in the interval } t \leqslant T \leqslant t + \delta t) \\
&\text{since what happens in the interval } t \text{ to} \\
&t + \delta t \text{ is independent of previous history} \\
= \quad &e^{-\alpha t} \times \alpha \delta t
\end{aligned}
$$

using (4.10) with $r = 0$ for $\Pr(Y = 0)$.

But by the definition of a p.d.f. (section 4.5.1) we have

$$
\Pr(t < T < t + \delta t) = f(t)\delta t
$$

Hence

$$
f(t) = \alpha e^{-\alpha t} \qquad 0 \leqslant t < \infty
$$

Thus the interval between events follows a negative exponential distribution with parameter α.

The mean interval is $1/\alpha$; this is the result which would be obtained if the events occurred at regular intervals at a mean rate of α per unit time.

Ex. 4.10 A long-lived radioactive source gives a mean count rate of 2 per second. If T seconds is the interval between consecutive emissions find (a) the mean and standard deviation of T, (b) $\Pr(0 < T < 0.5)$, (c) $\Pr(T > 1)$, (d) $\Pr(T > 1.5 | T > 0.5)$.

From the discussion above $f(t) = 2e^{-2t} \qquad 0 \le t < \infty$

and

(a) $\mu_T = \dfrac{1}{\alpha} = \dfrac{1}{2}, \qquad \sigma_T = \dfrac{1}{\alpha} = \dfrac{1}{2}$

(b) $\Pr(0 < T < 0.5) = \displaystyle\int_0^{0.5} 2e^{-2t}\,dt$ †

$\qquad\qquad\qquad = \left[-e^{-2t} \right]_0^{0.5}$

$\qquad\qquad\qquad = 1 - e^{-1}$

$\qquad\qquad\qquad = 0.6321$

(c) $\Pr(T > t) \qquad = \displaystyle\int_t^\infty 2e^{-2t}\,dt$

$\qquad\qquad\qquad = \left[-e^{-2t} \right]_t^\infty$

$\qquad\qquad\qquad = e^{-2t}$

$\qquad\qquad\qquad = e^{-2} = 0.1353 \quad \text{when } t = 1$

(d) $\Pr(T > 1.5 | T > 0.5) = \dfrac{\Pr\{(T > 1.5) \text{ and } (T > 0.5)\}}{\Pr(T > 0.5)}$

$\qquad\qquad\qquad = \dfrac{\Pr(T > 1.5)}{\Pr(T > 0.5)} \quad \text{since } T > 1.5 \text{ implies } T > 0.5$

$\qquad\qquad\qquad = \dfrac{e^{-2(1.5)}}{e^{-2(0.5)}} \quad \text{using (c) with } t = 0.5 \text{ and } 1.5$

$\qquad\qquad\qquad = e^{-2}$

$\qquad\qquad\qquad = \Pr(T > 1)$

† The required area under the p.d.f. curve is evaluated using integration. Readers not familiar with elementary calculus should omit Ex. 4.10.

The result obtained in part (d) may be generalised as follows

$$\Pr(t_2 + t_1 < T < t_3 + t_1 | T > t_1) = \Pr(t_2 < T < t_3) \text{ where } t_3 > t_2 \qquad (4.14)$$

Thus the conditional distribution of T given that $T > t_1$ is the same as the unconditional distribution of T. This is an illustration of what is sometimes referred to as the 'lack of memory' of the negative exponential distribution (compare the geometric distribution, section 4.3).

The sum of k interevent intervals, where the event occurrence has a Poisson distribution, follows a *gamma distribution* (see Suggestions for Further Reading). Thus the negative exponential distribution is a particular case of the gamma distribution with $k = 1$.

4.6 THE NORMAL DISTRIBUTION

Many continuous random variables encountered in practice follow, at least to a good approximation, a distribution called the *normal distribution* (also frequently referred to as the *Gaussian* or *error distribution*); these include

(1) Variations in automated processes; e.g. the weights of tablets made by a machine, the volume of a fluid dispensed by an automatic bottle filling machine, the weight of powder in machine-filled packets.
(2) Random experimental errors occurring in experiments in the physical sciences, e.g. chemistry, physics, astronomy.
(3) Variation of quantities examined in the life sciences, e.g. biology, agriculture.
(4) Even if the directly measured experimental variable does not follow a normal distribution, transformations can often be made to improve normality (e.g. by taking logarithms).
(5) Many other distributions tend to normality as sample sizes become larger.
(6) The distribution of the mean of a number of independent and identically distributed random variables generally approaches normality for large sample sizes whether or not the distribution of the individual variables is normal. (See section 5.2.1 for further discussion.)

The p.d.f. of a normally distributed random variable X has a rather complicated mathematical form, namely

$$f(x) = \frac{1}{(\sqrt{2\pi})\sigma} \exp\left[-\frac{1}{2\sigma^2}(x - \mu)^2\right] \qquad -\infty < x < \infty \qquad (4.15)$$

Using (4.15) it can be shown that:

(i)   the area bounded by the p.d.f. curve is unity
(ii) the mean of X is μ
(iii) the variance of X is σ^2.

The graph of $f(x)$ is a bell-shaped curve symmetrical about the mean ordinate $x = \mu$ (Fig. 4.2). It can be shown (see later) that almost all the area (0.9973) under the graph lies between the ordinates $x = \mu - 3\sigma$ and $x = \mu + 3\sigma$. This means that the normal distribution may be used to describe the behaviour of variables such as weight which are necessarily positive provided that $\mu - 3\sigma > 0$ so that practically all the area under the curve lies in the region $x > 0$.

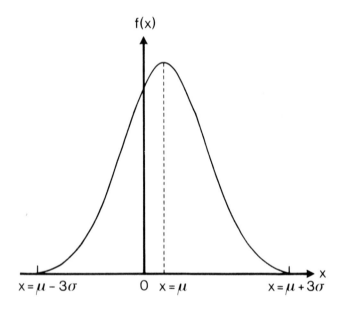

Fig. 4.2 – Probability density function for $X \sim N(\mu; \sigma^2)$.

If we know that a random variable X follows a normal distribution with given values of μ and σ^2 (we write $X \sim N(\mu; \sigma^2)$) then we can calculate

$$Pr(a < X < b)$$

as the area bounded by the curve $y = f(x)$, the x-axis and the ordinates $x = a$ and $x = b$ for any a or b. To evaluate this area we proceed as follows.

Let $Z = (X - \mu)/\sigma$. Then it can be shown that $Z \sim N(0; 1)$; that is Z is normally distributed with mean zero and variance 1. Z is called the *standard score*, or *standardised normal variable*, and the distribution of Z the *standard normal distribution*. The graph of the p.d.f. of Z is shown in Fig. 4.3.

Tables of the area under the standard normal distribution (either from $-\infty$ to z or from 0 to z) may be found in standard statistical tables. The area from $-\infty$ to z (for positive z) is tabulated in Table A1(a).

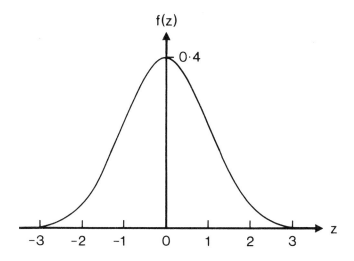

Fig. 4.3 – Probability density function for the standard normal variate.

Note that $\Pr(-\infty < Z < z)$ $=$ $\Pr(-\infty < Z < 0) + \Pr(0 < Z < z)$

$=$ $0.5 + \Pr(0 < Z < z)$

and

$$\Pr(-z < Z < 0) \quad = \quad \Pr(0 < Z < z),$$

both these results following from the symmetry of the graph about the vertical axis.

When $\quad X = a, Z = \dfrac{a - \mu}{\sigma} = z_1$

say, and when $X = b, Z = \dfrac{b - \mu}{\sigma} = z_2$

say, then $\quad \Pr(a < X < b) = \Pr(z_1 < Z < z_2)$

and the latter probability may be determined using Table A1(a).

Relatively cheap electronic calculators are now available which have the facility of working out areas under the standard normal curve. In some of the illustrative examples in this book such a facility has been used and the answers obtained will be slightly more accurate than those obtained using Table A1(a).

Ex. 4.11. Find (i) $\Pr(-0.46 < Z < 2.10)$ (ii) $\Pr(0.81 < Z < 1.94)$

(i) $\Pr(-0.46 < Z < 2.10) = \Pr(-0.46 < Z < 0) + \Pr(0 < Z < 2.10)$

$\qquad\qquad\qquad\qquad\quad = \Pr(0 < Z < 0.46) + \Pr(0 < Z < 2.10)$

$\qquad\qquad\qquad\qquad\quad = 0.6772 - 0.5 + 0.9821 - 0.5$

$\qquad\qquad\qquad\qquad\quad = 0.6593 \,.$

(ii) $\Pr(0.81 < Z < 1.94) \;=\; \Pr(-\infty < Z < 1.94) - \Pr(-\infty < Z < 0.81)$

$\qquad\qquad\qquad\qquad\quad = 0.9738 - 0.7910$

$\qquad\qquad\qquad\qquad\quad = 0.1828 \,.$

The method of obtaining the required probability is best determined by sketching the corresponding area on Fig. 4.3; for example in (ii) above draw the ordinates at $z = 0.81$ and $z = 1.94$ and shade the area between them.

Problem. Show that

(i) $\Pr(-1 < Z < 1) = 0.6827 \;(\doteq 2/3)$

(ii) $\Pr(-2 < Z < 2) = 0.9545 \;(\doteq 0.95)$

(iii) $\Pr(-3 < Z < 3) = 0.9973 \;(\doteq 1)$

Ex. 4.12. The lifetime of a certain type of light bulb is normally distributed about a mean of 700.0 hours with a standard deviation of 20.0 hours. Out of a random sample of 1000 bulbs how many bulbs would you expect to have lifetimes of (a) between 660.0 and 740.0 hours, (b) more than 750.0 hours, (c) 725 hours (to the nearest hour)?

Let X be the lifetime of a bulb. Then

$$Z = (X - \mu)/\sigma = (X - 700)/20$$

(a) $\qquad z_1 = \dfrac{660 - 700}{20} = -2; \quad z_2 = \dfrac{740 - 700}{20} = 2 \quad \text{and}$

$\Pr(660.0 < X < 740.0) \;=\; \Pr(-2 < Z < 2) \;=\; 2\Pr(0 < Z < 2) \;= 0.954$

Expected number of bulbs $= 1000 \times 0.954 = 954.$

(b) $\qquad z = \dfrac{750 - 700}{20} = 2.5 \text{ and}$

$\Pr(X > 750.0) = \Pr(Z > 2.5) = 1 - \Pr(Z < 2.5) = 1 - 0.9938 = 0.0062$

Expected number of bulbs $= 1000 \times 0.0062 \doteq 6.$

(c) Here we need the probability that X lies in the range 724.5 to 725.5 hours. Since this is a narrow range we shall find the values of z correct to three decimal places for greater accuracy in the final result.

$$z_1 = \frac{724.5 - 700}{20} = 1.225; z_2 = \frac{725.5 - 700}{20} = 1.275 \text{ and}$$

$$\Pr(724.5 < X < 725.5) = \Pr(1.225 < Z < 1.275)$$

$$= \Pr(Z < 1.275) - \Pr(Z < 1.225)$$

From Table A1(a) we find that $\Pr(Z < 1.27) = 0.8980$ and $\Pr(Z < 1.28) = 0.8997$, the difference between these two values being 0.0017. This difference corresponds to Z changing from 1.27 to 1.28, that is a change in Z of 0.01. Then the increment in probability corresponding to Z changing from 1.27 to 1.275 (a change of 0.005) will be $(0.005/0.01) \times 0.0017$ or 0.0009. Hence

$$\Pr(Z < 1.275) = 0.8980 + 0.0009 = 0.8989$$

Similarly $\Pr(Z < 1.225) = 0.8888 + \frac{1}{2}(0.8907 - 0.8888) = 0.8898$

Thus $\Pr(724.5 < X < 725.5) = 0.8989 - 0.8898 = 0.0091$

and the expected number of bulbs $= 1000 \times 0.0091 \doteq 9$.

The above method of finding $\Pr(Z < 1.275)$ and $\Pr(Z < 1.225)$, which cannot be read directly from the tables, is called *linear interpolation*.

4.6.1 Fitting a normal distribution to experimental data
Given a random sample of measurements we often need to determine if they can be regarded as a sample from a population with a given distribution. In particular to fit a normal distribution to a given set of data two approaches can be made:

(a) Use normal probability paper (section 4.6.1.1).
(b) Reduce the data to a frequency distribution, calculate the expected frequencies in each class under the assumption of normality (section 4.6.1.2).

4.6.1.1 Normal probability paper
One of the scales on *normal probability paper* is linear and the other is non-linear; the non-linear scale is such that if $X \sim N(\mu; \sigma^2)$ and we plot $F(x) = P(X \leqslant x)$ on the non-linear scale against x on the linear scale, a straight line will be obtained. $F(x)$ is called the *(cumulative) distribution of function* of X. Suppose for example we take $X \sim N(1;4)$. In Fig. 4.4, $F(x)$ is plotted against x for $-5 \leqslant x \leqslant 7$ on ordinary graph paper and, in Fig. 4.5, $F(x)$ is plotted against x on normal probability paper.

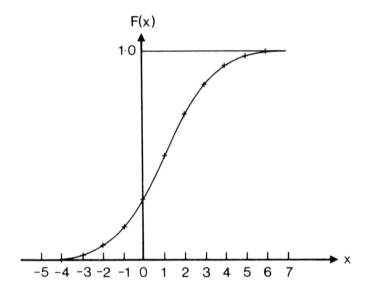

Fig. 4.4 – Plot of $F(x)$ against x, where $X \sim N(1; 4)$.

Suppose now that we have a sample of n observations arranged in ascending order of magnitude $x_1 \leqslant x_2 \leqslant \ldots \leqslant x_n$. The observed cumulative relative frequency, $\hat{F}(x_i)$, is then calculated from

$$\hat{F}(x_i) = \frac{\text{number of observations} \leqslant x_i}{n + 1}$$

$$= \frac{i}{n + 1}$$

Note that we divide by $n + 1$ rather than by n (compare the calculation of the percentiles of a sample, section 2.7.1).

We then plot the points $(x_i, \hat{F}(x_i))$ on the probability paper with the x-values plotted in the direction of the axis with the linear scale, choosing the scale so that the range $x_1, \ldots, x_n$ is included on the paper, and with the $\hat{F}(x_i)$ values plotted as percentages in the direction of the axis with the non-linear scale. If the sample is a random one from a normal distribution the points plotted in this way will lie approximately on a straight line. As you will see in the examples which follow, the mean and variance of the normal distribution may be deduced from the graph.

If the set of data is presented as a frequency distribution of n observations with class upper boundaries $y_1, \ldots, y_k$ we plot the points $(y_i, \hat{F}(y_i))$ on the paper where

$$\hat{F}(y_i) = \frac{\text{number of observations} \leqslant y_i}{n}$$

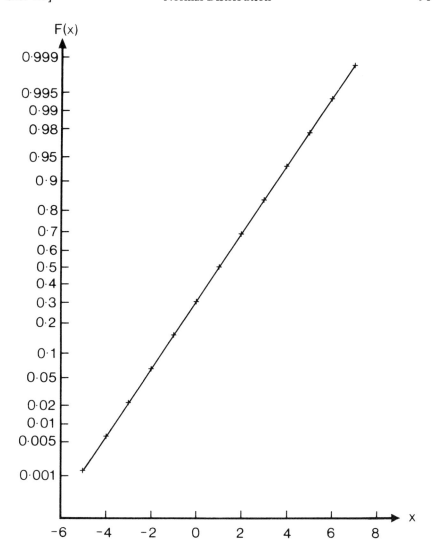

Fig. 4.5 – Plot of $F(x)$ against x, where $X \sim N(1;4)$, using normal probability paper.

This will give $\hat{F}(y_k) = 1$ and this is plotted as 0.9999 on the probability paper.

Ex. 4.13. The weights of 30 capsules (in mg) are given below

335.0,   338.0,   339.8,   341.0,   343.1,   343.5,   343.6,   346.7,   346.9,   347.9
348.5,   349.9,   350.9,   351.0,   351.5,   352.0,   352.3,   352.3,   352.5,   352.8
353.5,   354.1,   354.6,   354.8,   356.8,   357.0,   357.5,   358.6,   361.0,   366.0

Use normal probability paper to show that the data may be fitted by a normal
distribution and estimate the mean and standard deviation of the distribution.

The cumulative relative frequencies are tabulated below

i	x	$\hat{F}(x_i)$	i	x	$\hat{F}(x_i)$	i	x_i	$\hat{F}(x_i)$
1	335.0	0.032	11	348.5	0.355	21	353.5	0.677
2	338.0	0.065	12	349.9	0.387	22	354.1	0.710
3	339.8	0.097	13	350.9	0.419	23	354.6	0.742
4	341.0	0.129	14	351.0	0.452	24	354.8	0.774
5	343.1	0.161	15	351.5	0.484	25	356.8	0.806
6	343.5	0.194	16	352.0	0.516	26	357.0	0.839
7	343.6	0.226	17	352.3	0.548	27	357.5	0.871
8	346.7	0.258	18	352.3	0.581	28	358.6	0.903
9	346.9	0.290	19	352.5	0.613	29	361.0	0.935
10	347.9	0.323	20	352.8	0.645	30	366.0	0.968

· From the graph (Fig. 4.6) we see that the points lie approximately on a straight
line; hence it would appear that a normal distribution is a satisfactory fit. We

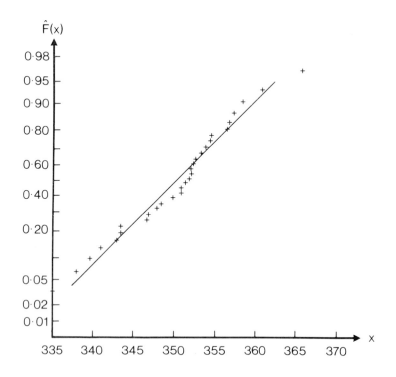

Fig. 4.6 — Fitting a normal distribution to the weights of capsules.

draw the best line through the points by eye. The one or two points at either end of the line are often unreliable and it is not important if the line does not pass near such points.

The mean and standard deviation of the distribution may be estimated from the graph as follows.

We have for the normal distribution

$$\Pr(X < \mu) = 0.5.$$

Thus we look for the point on the line with $\hat{F} = 0.5$ and the x-value of this point will give an estimate $\hat{\mu}$ of μ. From the graph we have

$$\hat{\mu} = 350.5.$$

We also have for the normal distribution

$$\Pr(X < \mu + \sigma) = \Pr(Z < 1) \doteq 0.84$$

Hence we look for the point on the line with $\hat{F} = 0.84$ and the x-value of this point will give an estimate of $\mu + \sigma$. From the graph we have

$$\hat{\mu} + \hat{\sigma} = 357.8$$

and since $\hat{\mu} = 350.5$, this leads to $\hat{\sigma} = 7.3$.

Ex. 4.14. The weights of 100 aspirin tablets are reduced to the following frequency distribution

Class mid-value (g)	0.3245	0.3265	0.3285	0.3305	0.3325	0.3345	0.3365	0.3385
Frequency	3	7	18	23	24	15	7	3

Using normal probability paper show that the data may be fitted by a normal distribution and estimate the mean and standard deviation of the distribution.

We need the class upper boundaries y_i and the corresponding cumulative relative frequencies $\hat{F}(y_i)$

x	0.3245	0.3265	0.3285	0.3305	0.3325	0.3345	0.3365	0.3385
y	0.3255	0.3275	0.3295	0.3315	0.3335	0.3355	0.3375	0.3395
f	3	7	18	23	24	15	7	3
cf	3	10	28	51	75	90	97	100
$\hat{F}(y)$	0.03	0.10	0.28	0.51	0.75	0.90	0.97	1.00

$\hat{F}(y)$ is plotted against y on probability paper in Fig. 4.7. From the graph we see that the points (omitting the last) lie on a straight line and hence we can assume that the data is well fitted by a normal distribution.

When $\hat{F} = 0.5, x = 0.3295 + 0.95 \times 0.002 = 0.3314$; hence $\hat{\mu} = 0.3314$.

When $\hat{F} = 0.84, x = 0.3335 + 0.52 \times 0.002 = 0.3345$; hence $\hat{\mu} + \hat{\sigma} = 0.3345$.

Therefore $\hat{\sigma} = 0.0031$.

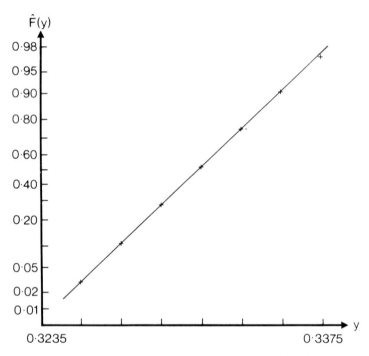

Fig. 4.7 – Fitting a normal distribution to the weights of 100 aspirin tablets.

4.6.1.2 Fitting a normal distribution to a frequency distribution

If there are sufficient observations to reduce the data to a frequency distribution, we can proceed as in the following example.

Ex. 4.15. The data of Table 2.1 may be reduced to the following distribution:

Class mid-value	0.3245	0.3265	0.3285	0.3305	0.3325	0.3345	0.3365	0.3385
Frequency	3	7	18	23	24	15	7	3

with mean $\bar{x} = 0.33142$ and standard deviation $s = 3.1581 \times 10^{-3}$.
Fit a normal distribution to the data and test for goodness of fit.

If we had not been given the mean and standard deviation they could of course have been calculated using the methods discussed earlier (sections 2.3 and 2.5). To fit a normal distribution to the data we proceed as follows; we first of all find the standard scores $z = (x - 0.33142)/0.0031581$ corresponding to the class boundaries. Suppose that a particular class has boundaries z_1 and z_2; the probability that a value of X falls into this class is $\Pr(z_1 < Z < z_2)$ and this probability, assuming a normal distribution, may be found from Table A1(a). So that all the area under the standard normal p.d.f. curve is included, we take the standard score of the lower boundary of the first class as $-\infty$, and that of the upper boundary of the final class as $+\infty$. To obtain the expected number of observations falling into a given class, we multiply the total number of observations (100 in this case) by the probability of an observation falling into the class. Visual comparison of the observed and expected frequencies indicates that the data are well fitted by the normal distribution. A statistical test for the goodness of fit will be developed in section 5.9.

Class boundaries x	Class boundaries z	Area from $-\infty$ to z[†]	Area in class[‡]	Expected frequency e	Observed frequency O
0.3235	$-\infty$	0.0000			
			0.0307	3.1	3
0.3255	-1.87	0.0307			
			0.0768	7.7	7
0.3275	-1.24	0.1075			
			0.1634	16.3	18
0.3295	-0.61	0.2709			
			0.2411	24.1	23
0.3315	0.03	0.5120			
			0.2334	23.3	24
0.3335	0.66	0.7454			
			0.1561	15.6	15
0.3355	1.29	0.9015			
			0.0717	7.2	7
0.3375	1.93	0.9732			
			0.0268	2.7	3
0.3395	∞	1.0000			

†Obtained from Table A1(a)
‡Obtained by differencing

4.7 USEFUL APPROXIMATIONS FOR THE BINOMIAL AND POISSON DISTRIBUTIONS

4.7.1 Poisson approximation to the binomial distribution

It can be shown that if n is large ($\geqslant 50$) and p is small ($p \leqslant 0.1$) with $np \leqslant 5$ then the binomial probabilities may be approximated by probabilities obtained from a Poisson distribution with the same mean ($\lambda = np$) as the binomial distribution (see equation (4.11) and Appendix 4.5).

The corresponding probabilities for $n = 50$ and $p = 0.1$ are tabulated in Table 4.1.

Table 4.1

n	0	1	2	3	4	5	6	7	8	9	$\geqslant 10$
Binomial	0.005	0.029	0.078	0.139	0.181	0.185	0.154	0.108	0.064	0.033	0.025
Poisson	0.007	0.034	0.084	0.140	0.175	0.175	0.146	0.104	0.065	0.036	0.032

The case $n = 100$ and $p = 0.05$ is illustrated in Table 4.2.

Table 4.2

n	0	1	2	3	4	5	6	7	8	9	$\geqslant 10$
Binomial	0.006	0.031	0.081	0.140	0.178	0.180	0.150	0.106	0.065	0.035	0.028
Poisson	0.007	0.034	0.084	0.140	0.175	0.175	0.146	0.104	0.065	0.036	0.032

It will be seen from the tables that the approximation improves (for a given value of $\lambda = np$) as n increases and p decreases, as would be expected.

Ex. 4.16. The probability of a person suffering an adverse reaction from an injection is 1/1000. Out of a random sample of 500 people find the probability that two or more persons suffer an adverse reaction.

Let X be the number of persons out of the sample of 500 who suffer an adverse reaction.

(a) *Binomial distribution*

$$X \sim b(500, 0.001)$$

$$
\begin{aligned}
\Pr(X \geqslant 2) &= 1 - \Pr(X = 0) - \Pr(X = 1) \\
&= 1 - 0.999^{500} - 500 \times 0.999^{499} \times 0.001 \\
&= 1 - 0.6064 - 0.3035 \\
&= 0.0901.
\end{aligned}
$$

(b) *Poisson distribution*
We can use the Poisson approximation to the binomial distribution, since $n = 500$ and $p = 0.001$ with $\lambda = np = 0.5$. Then

$$X \sim P(0.5)$$

and

$$
\begin{aligned}
\Pr(X \geq 2) &= 1 - \Pr(X = 0) - \Pr(X = 1) \\
&= 1 - e^{-0.5} - 0.5e^{-0.5} \\
&= 1 - 0.6065 - 0.3033 \\
&= 0.0902 .
\end{aligned}
$$

Here the approximation is excellent; there is slightly less work involved in working out the probabilities using the Poisson approximation.

4.7.2 Normal approximation to the binomial distribution
It can be shown that if np and nq are both greater than or equal to about 5, the binomial distribution $b(n, p)$ may be approximated by the normal distribution $N(np; npq)$; that is the binomial distribution may be approximated by a normal distribution with the same mean and variance. As the binomial distribution is discrete and the normal continuous, a *continuity correction* must be applied in using the approximation. The method of doing this is illustrated in the examples below.

Ex. 4.17. Find the probability distribution of the number X of heads in 10 tosses of a fair coin.

(a) *Binomial distribution*

$$
\text{Using } \Pr(X = r) = \binom{10}{r} \left(\frac{1}{2}\right)^{10-r} \left(\frac{1}{2}\right)^{r} = \binom{10}{r} \frac{1}{2^{10}} \text{ leads to}
$$

X	0	1	2	3	4	5	6	7	8	9	10
$\Pr(X = r)$	0.001	0.010	0.044	0.117	0.205	0.246	0.205	0.117	0.044	0.010	0.001

These probabilities are illustrated in the bar diagram of Fig. 4.8.

(b) *Normal approximation to the binomial distribution*
Here $np = nq = 5$ and the conditions for using the normal approximation $(np \geq 5, nq \geq 5)$ are just satisfied.
    Let Y be the normal variable approximating to X. Then

$$Y \sim N(\mu; \sigma^2)$$

where $\mu = np = 10 \times \dfrac{1}{2} = 5$ and $\sigma^2 = npq = 10 \times \dfrac{1}{2} \times \dfrac{1}{2} = 2.5$.

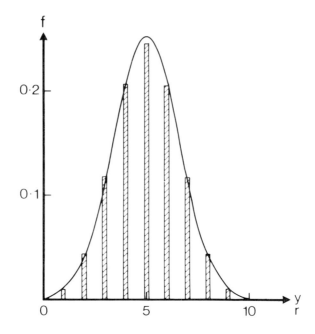

Fig. 4.8 – Normal approximation to the binomial distribution.

The p.d.f. of Y is illustrated in Fig. 4.8. In order to approximate the area under the p.d.f. curve we must take

$$\Pr(X = r) = \Pr\left(r - \frac{1}{2} < Y < r + \frac{1}{2}\right)$$

Then for example

$$\Pr(3 \leqslant X \leqslant 6) = \Pr\left(3 - \frac{1}{2} < Y < 6 + \frac{1}{2}\right) = \Pr(2.5 < Y < 6.5)$$

and

$$\Pr(2 < X < 7) = \Pr\left(2 + \frac{1}{2} < Y < 7 - \frac{1}{2}\right) = \Pr(2.5 < Y < 6.5),$$

where these last two probabilities are of course the same.

The direction in which the continuity correction has to be applied is most easily obtained from a sketch similar to that of Fig. 4.8.

Using

$$\Pr(X = r) = \Pr\left(r - \frac{1}{2} < Y < r + \frac{1}{2}\right) = \Pr\left(\frac{r - \frac{1}{2} - 5}{\sqrt{2.5}} < Z < \frac{r + \frac{1}{2} - 5}{\sqrt{2.5}}\right)$$

where $Z \sim N(0; 1)$ leads to the following table of probabilities

X	0	1	2	3	4	5	6	7	8	9	10
$\Pr(X = r)$	0.002	0.011	0.043	0.114	0.205	0.248	0.205	0.114	0.043	0.011	0.002

The results obtained are in close agreement with those obtained in part (a). Note however that the relative accuracy of the approximation is poorer in the tails of the distribution.

Ex. 4.18. A machine is producing bolts of which a proportion $p = 0.05$ are defective. A random sample of 400 bolts is taken from a large batch and the number of defective bolts D in the sample noted. Find (i) $\Pr(D = 10)$, (ii) $\Pr(D = 20)$, (iii) $\Pr(D = 30)$.

If the batch is accepted if $D \leqslant 30$ and rejected if $D > 30$ find the probability of (a) accepting a batch with a proportion $p = 0.05$ of defectives, (b) rejecting a batch with a proportion $p = 0.10$ of defectives.

(a) *Binomial distribution*

(i) $\Pr(D = 10) = \binom{400}{10} (0.95)^{390} (0.05)^{10} = 0.0052$.

(ii) $\Pr(D = 20) = \binom{400}{20} (0.95)^{380} (0.05)^{20} = 0.0911$.

(iii) $\Pr(D = 30) = \binom{400}{30} (0.95)^{370} (0.05)^{30} = 0.0076$.

A desk-top computer program gives

$$\Pr(\text{accepting a batch} | p = 0.05) = \Pr(D \leqslant 30) = 0.989.$$

and

$$\Pr(\text{rejecting a batch} | p = 0.1) = 1 - \Pr(D \leqslant 30) = 1 - 0.0524$$

$$= 0.948.$$

(b) *Normal distribution*

Since $np = 20, nq = 380$ we may approximate the distribution of D by $Y \sim N(20; 19)$

(i) $\Pr(D = 10) = \Pr(9.5 < Y < 10.5)$

$\qquad\qquad\qquad = \Pr(-2.4089 < Z < -2.1794)$

$\qquad\qquad\qquad = 0.0066$.

(ii) $\Pr(D = 20) = \Pr(19.5 < Y < 20.5)$

$\qquad\qquad\qquad = \Pr(-0.1147 < Z < 0.1147)$

$\qquad\qquad\qquad = 0.0913$.

(iii) $\Pr(D = 30) = \Pr(29.5 < Y < 30.5)$

$$= \Pr(2.1794 < Z < 2.4089)$$

$$= 0.0066.$$

$\Pr(\text{accepting a batch} | p = 0.05) = \Pr(Y < 30.5)$

$$= \Pr(Z < 2.4089)$$

$$= 0.992,$$

and

$\Pr(\text{rejecting a batch} | p = 0.1) = 1 - \Pr(U < 30.5), \text{ where } U \sim N(40; 36)$

$$= 1 - \Pr(Z < -1.5833)$$

$$= 0.943.$$

The two sets of results are in close agreement, but the working using the normal approximation is clearly more straightforward.

4.7.3 Normal approximation to the Poisson distribution

For $\lambda > 5$ the Poisson distribution $X \sim P(\lambda)$ may be approximated by $Y \sim N(\lambda; \lambda)$, with a continuity correction being applied as in the case of the normal approximation to the binomial distribution. The approximation improves as λ increases as illustrated in Table 4.3 ($\lambda = 5$) and Table 4.4 ($\lambda = 10$). As with the normal approximation to the binomial distribution it will be observed that the relative error is larger in the tails of the distribution than for values near the mean.

Table 4.3
Normal approximation to Poisson ($\lambda = 5$)

X	0	1	2	3	4	5	6	7	8	9	10	11
Poisson	0.007	0.034	0.084	0.140	0.175	0.175	0.146	0.104	0.065	0.036	0.018	0.008
Normal	0.015	0.037	0.073	0.119	0.160	0.177	0.160	0.119	0.073	0.037	0.015	0.005

Table 4.4
Normal approximation to Poisson ($\lambda = 10$)

X	0	1	2	3	4	5	6	7	8	9	10
Poisson	0.000	0.000	0.002	0.008	0.019	0.038	0.063	0.090	0.113	0.125	0.125
Normal	0.001	0.002	0.005	0.011	0.021	0.036	0.057	0.080	0.103	0.120	0.126

X	11	12	13	14	15	16	17	18	19	20
Poisson	0.114	0.095	0.073	0.052	0.035	0.022	0.013	0.007	0.004	0.002
Normal	0.120	0.103	0.080	0.057	0.036	0.021	0.011	0.005	0.002	0.001

4.8 BASIC PROGRAMS FOR THE GENERATION OF DATA FROM THE BINOMIAL, POISSON AND NORMAL DISTRIBUTIONS

4.8.1 Binomial distribution

Program 4.1 generates observations according to $X \sim b(n, p)$, where the values of n and p are input by the user. The distribution mean and variance are calculated from $\mu = np$ (4.5) and $\sigma^2 = npq$ (4.6). The theoretical probabilities are generated according to the algorithm (4.7).

Each observation X is obtained by generating n random numbers uniformly distributed over the range 0 to 1. If a random number is less than or equal to p a 'success' is scored otherwise a 'failure' is scored. X is then simply the number of successes out of the n random numbers (or trials).

4.8.2 Poisson distribution

Program 4.2 generates observations according to $X \sim P(\lambda)$, where the value λ is input by the user. The distribution mean and variance are calculated from $\mu = \sigma^2 = \lambda$. The theoretical probabilities are generated according to the algorithm (4.12).

The observations from the distribution are generated according to the principle illustrated in Fig. 3.4. The lengths of the segments OA_1, $A_1 A_2$, . . . represent $\Pr(X = 0)$, $\Pr(X = 1)$, . . . Random numbers uniformly distributed over the range 0 to 1 are generated. If the number falls in the range OA_1 a value $X = 0$ is generated; if it falls in the range $A_1 A_2$ a value $X = 1$ is generated, and so on. In theory

$$\sum_{r=0}^{\infty} \Pr(X = r) = 1$$

so that in practice a reasonable upper bound r_u must be chosen for r. In the program this bound is chosen such that

$$\sum_{r=0}^{r_u} \Pr(X = r) = 0.99999$$

and any random numbers lying in the range 0.99999 to 1 are not used.

4.8.3 Normal distribution

Program 4.3 generates observations according to $X \sim N(\mu; \sigma^2)$ where the values μ and σ are input by the user. A convenient lower bound LB, less than or equal to $\mu - 4\sigma$, and the class range R are input by the user. The program then sets the number of classes NC so that the range LB to $\mu + 4\sigma$ is covered by the classes. It is highly unlikely that an observation X will be generated which lies outside this range.

The theoretical probabilities are calculated by evaluating the integral

$$\Pr(z_1 < Z < z_2) = \frac{1}{\sqrt{2\pi}} \int_{z_1}^{z_2} \exp(-z^2/2)dz$$

using Simpson's rule with an interval $h = 0.005$. Here z_1 and z_2 are the standard scores corresponding to the lower and upper bounds of the class concerned.

The normally distributed observations X are generated using the concept that the sampling distribution of the mean $\overline{Y}$ of n observations, where n is greater than 30, is generally normal irrespective of whether Y itself is normally distributed or not. This is perhaps not the simplest way of obtaining normally distributed observations, but it is a useful illustration of the concept of the sampling distribution of the mean which is discussed more fully in Chapter 5. Details of the algorithm used are given in Appendix 4.6.

Program listings and sample outputs
(a) Program 4.1

```
100 PRINT"☐":REM**CLEAR SCREEN
110 PRINT"            **PROGRAM 4.1**"
120 PRINT"      GENERATION OF DATA FROM B(N,P)
130 PRINT"ESTIMATES OF MEAN VARIANCE,PROBABILITIES"
140 PRINT
150 DIM PR(20),PH(20),F(20)
†160 PRINT"N=";:INPUT N
170 PRINT"P=";:INPUT P
180 PRINT"PRINT-OUT INTERVAL=";:INPUT KP
190 XO=RND(-2):REM**INITIALISE RANDOM NUMBER SEQUENCE
200 REM**ZERO FREQUENCIES
210 FOR I=0 TO N : F(I)=0 : NEXT I
220 REM**CALCULATION OF THEOR.PROBS. PR(I),
230 REM**MEAN M, AND VARIANCE V
240 Q=1-P : R=P/Q
250 PR(0)=Q↑N
260 FOR I=0 TO N-1
270 PR(I+1)=(N-I)/(I+1)*R*PR(I)
280 NEXT I
290 M=N*P : V=N*P*Q
300 REM**ZERO SUM S1X AND SUM OF SQUARES
310 REM**S2X2 OF THE OBSERVATIONS X
320 S1X=0 : S2X2=0
330 REM**GENERATE OBSERVATIONS X.
340 REM**NC IS THE CURRENT NO. OF OBS.
350 NC=0
360 NC=NC+1
370 X=0
380 FOR I=1 TO N
390 U=RND(1)
400 IF U<=P THEN X=X+1
410 NEXT I
420 REM**UPDATE FREQ.DIST.,SUM AND SUM OF SQUARES
430 F(X)=F(X)+1
440 S1X=S1X+X
450 S2X2=S2X2+X*X
460 IF NC<>INT(NC/KP)*KP THEN 360
```

```
470 REM**CALCULATE PROB.,MEAN AND VAR. ESTS.
480 MH=INT(1000*S1X/NC+0.5)/1000
490 VH=(S2X2-S1X↑2/NC)/(NC-1)
500 VH=INT(1000*VH+0.5)/1000
510 FOR I=0 TO N
520 PH(I)=F(I)/NC
530 NEXT I
540 PRINT"❑":REM**CLEAR SCREEN
550 PRINT"DISTRIBUTION IS B(";N;",";P;")"
560 PRINT
570 PRINT"NUMBER OF OBSERVATIONS=";NC
580 PRINT"MEAN=";M;TAB(16);"ESTIMATED MEAN=";MH
590 PRINT"VARIANCE=";V;TAB(16);"EST. VARIANCE=";VH
600 PRINT
610 PRINT"PRESS ANY KEY TO CONTINUE"
620 GET A$ : IF A$="" THEN 620
630 PRINT"❑":REM**CLEAR SCREEN
640 PRINT"   X  THEORETICAL    ESTIMATED"
650 PRINT"       PROB.          PROB."
660 FOR I=0 TO N
670 E=INT(1000*PR(I)+0.5)/1000
680 F=INT(1000*PH(I)+0.5)/1000
690 PRINT TAB(2);I;TAB(8);E;TAB(21);F
700 NEXT I
710 PRINT
720 PRINT"DO YOU WANT A FURTHER";KP;" OBSERVATIONS (Y/N)";
730 INPUT A$
740 IF A$="Y" THEN 360
750 END
```

†Alternatively,
 160 INPUT "N ="; N
 is acceptable in several versions of BASIC

Sample of the output from program 4.1

```
DISTRIBUTION IS B( 8 , .2 )

NUMBER OF OBSERVATIONS= 200
MEAN=      1.6          ESTIMATED MEAN= 1.455
VARIANCE= 1.28          EST. VARIANCE=  1.214

   X   THEORETICAL      ESTIMATED
        PROB.            PROB.
   0    0.168           0.220
   1    0.336           0.325
   2    0.294           0.275
   3    0.147           0.140
   4    0.046           0.040
   5    0.009           0.000
   6    0.001           0.000
   7    0.000           0.000
   8    0.000           0.000
```

```
DISTRIBUTION IS B( 8 , .2 )

NUMBER OF OBSERVATIONS= 400
MEAN=      1.6            ESTIMATED MEAN= 1.525
VARIANCE= 1.28           EST. VARIANCE=  1.172

    X    THEORETICAL      ESTIMATED
           PROB.            PROB.
    0      0.168            0.175
    1      0.336            0.350
    2      0.294            0.300
    3      0.147            0.135
    4      0.046            0.030
    5      0.009            0.010
    6      0.001            0.000
    7      0.000            0.000
    8      0.000            0.000


DISTRIBUTION IS B( 8 , .2 )

NUMBER OF OBSERVATIONS= 600
MEAN=      1.6            ESTIMATED MEAN= 1.57
VARIANCE= 1.28           EST. VARIANCE=  1.194

    X    THEORETICAL      ESTIMATED
           PROB.            PROB.
    0      0.168            0.157
    1      0.336            0.363
    2      0.294            0.292
    3      0.147            0.142
    4      0.046            0.035
    5      0.009            0.012
    6      0.001            0.000
    7      0.000            0.000
    8      0.000            0.000
```

(b) Program 4.2

```
100 PRINT"▯":REM**CLEAR SCREEN
110 PRINT"           **PROGRAM 4.2**
120 PRINT" GENERATION OF DATA FROM A POISSON"
130 PRINT"   DISTRIBUTION WITH MEAN LAMBDA"
140 PRINT"AND ESTS. OF THE MEAN,VAR. AND PROBS."
150 DIM P(50),PH(50),F(50),CP(50)
160 PRINT
170 PRINT"LAMBDA=";:INPUT L
180 PRINT"PRINT-OUT INTERVAL=";:INPUT KP
190 X0=RND(-2):REM**INITIALISE RANDOM NUMBER SEQUENCE
200 REM**CALCULATION OF THEOR.PROBS.,P(I)
210 REM**CUMULATIVE PROBS.,CP(I)
220 REM**MEAN M, AND VARIANCE V
230 M=L : V=L
240 P(0)=EXP(-L)
250 CP(0)=P(0)
260 I=-1
```

```
270 I=I+1
280 P(I+1)=P(I)*L/(I+1)
290 CP(I+1)=CP(I)+P(I+1)
300 REM**STOP IF CUM.PROB. CLOSE TO 1
310 IF 1-CP(I+1)<1E-05 THEN 330
320 GOTO 270
330 NP=I
340 REM**GENERATE OBSERVATIONS X
350 REM**NC IS CURRENT COUNT
360 REM**ZERO FREQUENCIES
370 FOR I=0 TO NP:F(I)=0:NEXT I
380 S1X=0 : S2X2=0
390 NC=0
400 NC=NC+1
410 U=RND(1) : IF U>0.99999 THEN 410
420 X=-1
430 X=X+1
440 IF U<=CP(X) THEN 460
450 GOTO 430
460 F(X)=F(X)+1
470 S1X=S1X+X
480 S2X2=S2X2+X*X
490 IF NC<>INT(NC/KP)*KP THEN 400
500 REM**CALCULATE PROB., MEAN AND VAR. ESTS.
510 MH=INT(1000*S1X/NC+0.5)/1000
520 VH=(S2X2-S1X↑2/NC)/(NC-1)
530 VH=INT(1000*VH+0.5)/1000
540 FOR I=0 TO NP
550 PH(I)=F(I)/NC
560 NEXT I
570 PRINT"▯":REM**CLEAR SCREEN
580 PRINT
590 PRINT"DISTRIBUTION IS P(";L;")"
600 PRINT
610 PRINT"NUMBER OF OBSERVATIONS=";NC
620 PRINT"MEAN=";M,"ESTIMATED MEAN=";MH
630 PRINT"VARIANCE=";V,"EST. VARIANCE=";VH
640 PRINT"PRESS ANY KEY TO CONTINUE"
650 GET A$ : IF A$="" THEN 650
660 PRINT"▯":REM**CLEAR SCREEN
670 PRINT"   X   THEORETICAL    ESTIMATED"
680 PRINT"          PROB.        PROB."
690 FOR I=0 TO NP
700 E=INT(1000*P(I)+0.5)/1000
710 F=INT(1000*PH(I)+0.5)/1000
720 PRINT TAB(2);I;TAB(8);E;TAB(21);F
730 NEXT I
740 PRINT
750 PRINT"DO YOU WANT A FURTHER ";KP;" OBSERVATIONS (Y/N)";
760 INPUT A$
770 IF A$="Y" THEN 400
780 END
```

Sample of the output from program 4.2

```
DISTRIBUTION IS P( 2 )

NUMBER OF OBSERVATIONS= 200
MEAN= 2                    ESTIMATED MEAN= 1.865
VARIANCE= 2                EST. VARIANCE= 1.846

    X   THEORETICAL   ESTIMATED
           PROB.         PROB.
    0      0.135         0.135
    1      0.271         0.330
    2      0.271         0.235
    3      0.180         0.195
    4      0.090         0.065
    5      0.036         0.030
    6      0.012         0.005
    7      0.003         0.000
    8      0.001         0.005
    9      0.000         0.000


DISTRIBUTION IS P( 2 )

NUMBER OF OBSERVATIONS= 400
MEAN= 2                    ESTIMATED MEAN= 1.995
VARIANCE= 2                EST. VARIANCE= 2.03

    X   THEORETICAL   ESTIMATED
           PROB.         PROB.
    0      0.135         0.125
    1      0.271         0.295
    2      0.271         0.258
    3      0.180         0.193
    4      0.090         0.073
    5      0.036         0.035
    6      0.012         0.018
    7      0.003         0.003
    8      0.001         0.003
    9      0.000         0.000


DISTRIBUTION IS P( 2 )

NUMBER OF OBSERVATIONS= 600
MEAN= 2                    ESTIMATED MEAN= 2.037
VARIANCE= 2                EST. VARIANCE= 2.059

    X   THEORETICAL   ESTIMATED
           PROB.         PROB.
    0      0.135         0.120
    1      0.271         0.278
    2      0.271         0.277
    3      0.180         0.190
    4      0.090         0.075
    5      0.036         0.035
    6      0.012         0.017
    7      0.003         0.007
    8      0.001         0.002
    9      0.000         0.000
```

(c) Program 4.3

```
100 PRINT"⬛":REM**CLEAR SCREEN
110 PRINT"       **PROGRAM 4.3**"
120 PRINT" GENERATION OF DATA FROM N(M;S↑2)"
130 PRINT"ESTS. OF PROBS.,MEAN AND VARIANCE"
140 PRINT
150 DIM P(50),F(50),PH(50)
160 PRINT"M=";:INPUT M
170 PRINT"S=";:INPUT S
180 PRINT"PRINT-OUT INTERVAL=";:INPUT KP
190 REM**CALCULATION OF THEORETICAL PROBABILITIES
200 PRINT"LOWER BOUND(<=M-4*S)=";:INPUT LB
210 PRINT"CLASS WIDTH=";:INPUT R
220 NC=INT((M+4*S-LB)/R)+1
230 PI=3.1415927
240 H=0.005
250 N=INT(R/(2*H*S)+0.00001)
260 BL=LB
270 FOR I=1 TO NC
280 BU=BL+R
290 REM**CALC. STANDARD SCORES FOR CLASS BOUNDARIES
300 Z1=(BL-M)/S
310 Z2=(BU-M)/S
315 REM**CALCULATE AREA IN CLASS
320 GOSUB 920
330 REM**P(I) IS PROBABILITY FOR CLASS
340 P(I)=P
350 BL=BU
360 NEXT I
370 REM**
380 REM**GENERATION OF OBSERVATIONS X,
390 REM**REDUCTION TO A FREQUENCY DIST.
400 REM**ESTIMATES OF PROB.,MEAN AND VARIANCE
410 X0=RND(-2):REM**INITIALISE RANDOM NUMBER SEQUENCE
420 REM**ZERO FREQUENCIES AND SUMS
430 FOR I=1 TO NC : F(I)=0 : NEXT I
440 S1X=0 : S2X2=0
450 REM**IC*KP IS CURRENT COUNT
460 IC=0
470 IC=IC+1
480 FOR I=1 TO KP
490 X=0
500 FOR J=1 TO 40
510 X=X+RND(1)
520 NEXT J
530 X=M+S*(X-20)/SQR(10/3)
540 REM**IGNORE ANY X>4*S FROM M
550 IF X<M-4*S OR X>M+4*S THEN 490
560 S1X=S1X+X
570 S2X2=S2X2+X*X
580 REM**ASSIGN X TO ITS CLASS
590 K=INT((X-LB)/R)+1
600 F(K)=F(K)+1
610 NEXT I
620 NT=IC*KP
630 MH=INT(1000*S1X/NT+0.5)/1000
640 VH=(S2X2-S1X↑2/NT)/(NT-1)
650 VH=INT(1000*VH+0.5)/1000
660 FOR I=1 TO NC
670 PH(I)=F(I)/NT
680 NEXT I
```

```
690 PRINT"[]":REM**CLEAR SCREEN
700 PRINT
710 PRINT"DISTRIBUTION IS N(";M;",";St2;")"
720 PRINT"NUMBER OF OBSERVATIONS=";NT
730 PRINT"MEAN=";M,"ESTIMATED MEAN=";MH
740 PRINT"VARIANCE=";St2,"EST.VAR.=";VH
750 PRINT"PRESS ANY KEY TO CONTINUE"
760 GET A$ : IF A$=""THEN 760
770 PRINT"[]":REM**CLEAR SCREEN
780 PRINT
790 PRINT"  CLASS      THEORETICAL   ESTIMATED"
800 PRINT" MID-VALUE      PROB.        PROB."
810 REM**CALCULATE MID-VALUE OF FIRST CLASS
820 MV=INT(100*(LB+R/2)+0.5)/100
830 FOR I=1 TO NC
840 E=INT(1000*P(I)+0.5)/1000
850 F=INT(1000*PH(I)+0.5)/1000
860 PRINT TAB(2);MV;TAB(15);E;TAB(31);F
865 MV=INT(100*(MV+R)+0.5)/100
870 NEXT I
880 PRINT"DO YOU WANT A FURTHER ";KP;" OBSERVATIONS (Y/N)";
890 INPUT A$
900 IF A$="Y" THEN 470
910 END
920 REM**CALCULATION OF AREA UNDER STANDARD NORMAL CURVE
930 REM**USING SIMPSON'S RULE
940 SO=0 : SE=0
950 FOR I1=1 TO N
960 SO=SO+EXP(-(Z1+(2*I1-1)*H)t2/2)
970 NEXT I1
980 FOR I1=1 TO N-1
990 SE=SE+EXP(-(Z1+2*I1*H)t2/2)
1000 NEXT I1
1010 C=SQR(2*PI)
1020 P=(EXP(-Z1t2/2)+EXP(-(Z1+2*N*H)t2/2)+4*SO+2*SE)*H/(3*C)
1030 REM**CORRECTION FOR AREA AT END OF CLASS
1040 P=P+(EXP(-(Z1+2*N*H)t2/2)+EXP(-Z2t2/2))*(Z2-Z1-2*N*H)/(2*C)
1050 RETURN
```

Sample of the output from program 4.3

```
DISTRIBUTION IS N( 1 , 4 )

NUMBER OF OBSERVATIONS= 800
MEAN= 1                        ESTIMATED MEAN= 1.183
VARIANCE= 4                    EST.VAR.= 4.109

     CLASS       THEORETICAL    ESTIMATED
   MID-VALUE        PROB.         PROB.
     -6.5          0.000         0.001
     -5.5          0.001         0.000
     -4.5          0.005         0.003
     -3.5          0.017         0.016
     -2.5          0.044         0.048
     -1.5          0.092         0.074
     -.5           0.150         0.139
      .5           0.191         0.173
     1.5           0.191         0.208
     2.5           0.150         0.155
     3.5           0.092         0.109
     4.5           0.044         0.050
     5.5           0.017         0.021
     6.5           0.005         0.004
     7.5           0.001         0.001
     8.5           0.000         0.000
     9.5           0.000         0.000
```

APPENDIX 4.1 DERIVATION OF BINOMIAL PROBABILITIES

Let $X \sim b(n, p)$; then $\Pr(X = r)$ is the probability of obtaining r successes (and hence $n - r$ failures) in n independent trials, with the probability of success, p, remaining constant from trial to trial. The probability of obtaining r successes and $n - r$ failures in a fixed sequence is

$$p \times p \ldots (r \text{ times}) \times q \times q \ldots (n - r \text{ times})$$

$$= p^r q^{n-r}$$

But the r successes can occur in $\binom{n}{r}$ ways, which is the number of ways in which r trials can be chosen from the n trials in the experiment, and each of these ways is equally probable. Then

$$\Pr(X = r) = \binom{n}{r} q^{n-r} p^r \tag{4.4}$$

The cases $r = 0$ and $r = n$ may be included in this formula using $\binom{n}{0} - \binom{n}{n} - 1$.

APPENDIX 4.2 THE MEAN AND VARIANCE OF THE BINOMIAL DISTRIBUTION

Let $X \sim b(n, p)$. Then

$$\mu = E(X) = \sum_{r=0}^{n} r \times \Pr(X = r)$$

$$= \sum_{r=1}^{n} r \times \frac{n!}{r!(n-r)!} q^{n-r} p^r \quad \text{since the term with } r = 0 \text{ is zero}$$

$$= \sum_{r=1}^{n} \frac{n!}{(r-1)!(n-r)!} q^{n-r} p^r$$

$$= \sum_{s=0}^{n-1} \frac{n!}{s!(n-s-1)!} q^{n-s-1} p^{s+1} \quad \text{substituting } r = s + 1$$

$$= np \sum_{s=0}^{n-1} \frac{(n-1)!}{s!(n-1-s)!} q^{n-1-s} p^s$$

$$= np \sum_{s=0}^{n-1} \binom{n-1}{s} q^{n-1-s} p^s$$

$$= np(q + p)^{n-1} \quad \text{from the binomial expansion of } (q + p)^{n-1}$$

$$= np \qquad\qquad\qquad\qquad\qquad\qquad\qquad\qquad (4.5)$$

using $(q + p) = 1$.

To evaluate the variance we first extend the idea of expected values. Let X be a discrete random variable with possible values $x_1, x_2, \ldots, x_k$. Then the expected value $E(g(X))$ of a function $g(X)$ of X is defined by

$$E(g(X)) = \sum_{i=1}^{k} g(x_i) \Pr(X = x_i)$$

In terms of expected values the variance σ^2 of the distribution can be written as

$$\sigma^2 = E(X^2) - (E(X))^2$$

To find the variance for the binomial distribution it is simpler to first evaluate $E(X(X - 1))$. Then

$$E(X(X - 1)) = \sum_{r=0}^{n} r(r - 1) \times \Pr(X = r)$$

$$= \sum_{r=2}^{n} r(r - 1) \frac{n!}{r!(n-r)!} q^{n-r} p^r \quad \begin{array}{l} \text{since the terms with } r = 0 \\ \text{and } r = 1 \text{ are zero} \end{array}$$

$$= \sum_{r=2}^{n} \frac{n!}{(r-2)!(n-r)!} \; q^{n-r} p^r$$

$$= \sum_{s=0}^{n-2} \frac{n!}{s!(n-s-2)!} \; q^{n-s-2} p^{s+2} \quad \text{substituting } r = s+2$$

$$= n(n-1)p^2 \sum_{s=0}^{n-2} \frac{(n-2)!}{s!(n-2-s)!} \; q^{n-2-s} p^s$$

$$= n(n-1)p^2 \sum_{s=0}^{n-2} \binom{n-2}{s} q^{n-2-s} p^s$$

$$= n(n-1)p^2 \, (q+p)^{n-2}$$

$$= n(n-1)p^2$$

using $q + p = 1$.

Therefore $\mathrm{E}(X^2) = n(n-1)p^2 + \mathrm{E}(X) = n(n-1)p^2 + np$

and
$$\begin{aligned}
\sigma^2 &= \mathrm{E}(X^2) - (\mathrm{E}(X))^2 \\
&= n(n-1)p^2 + np - n^2 p^2 \\
&= np(1-p) \\
&= npq.
\end{aligned} \tag{4.6}$$

APPENDIX 4.3 THE MEAN AND VARIANCE OF THE POISSON DISTRIBUTION

Let $X \sim P(\lambda)$

$$\mu = \mathrm{E}(X) = \sum_{r=0}^{\infty} r \times \mathrm{Pr}(X = r)$$

$$= \sum_{r=0}^{\infty} r e^{-\lambda} \frac{\lambda^r}{r!}$$

$$= e^{-\lambda} \sum_{r=1}^{\infty} \frac{\lambda^r}{(r-1)!}$$

$$= e^{-\lambda} \sum_{s=0}^{\infty} \frac{\lambda^{s+1}}{s!} \quad \text{substituting } r = s+1$$

$$= \lambda e^{-\lambda} \sum_{s=0}^{\infty} \frac{\lambda^s}{s!}$$

$$= \lambda e^{-\lambda} \times e^{\lambda}$$

$$= \lambda .$$

To find an expression for the variance we first find $E[X(X-1)]$

$$E(X(X-1)) = \sum_{r=0}^{\infty} e^{-\lambda} \frac{\lambda^r}{r!} r(r-1)$$

$$= \sum_{r=2}^{\infty} e^{-\lambda} \frac{\lambda^r}{(r-2)!}$$

$$= e^{-\lambda} \sum_{s=0}^{\infty} \frac{\lambda^{s+2}}{s!} \quad \text{letting } r = s + 2$$

$$= \lambda^2 e^{-\lambda} \sum_{s=0}^{\infty} \frac{\lambda^s}{s!}$$

$$= \lambda^2 e^{-\lambda} \times e^{\lambda}$$

$$= \lambda^2$$

therefore $\quad E(X^2) = \lambda^2 + E(X) = \lambda^2 + \lambda$

and $\qquad \sigma^2 = E(X^2) - (E(X))^2 = \lambda^2 + \lambda - \lambda^2 = \lambda .$

APPENDIX 4.4 THE MEAN AND VARIANCE OF THE NEGATIVE EXPONENTIAL DISTRIBUTION

For a continuous random variable X with p.d.f. $f(x)$, $-\infty < x < \infty$, the mean μ (or expected value $E(X)$) of X is given by the formula

$$\mu = E(X) = \int_{-\infty}^{\infty} x f(x) \, dx .$$

Then the variance σ^2 of X is given by

$$\sigma^2 = E(X^2) - (E(X))^2$$

$$= \int_{-\infty}^{\infty} x^2 f(x) dx - \mu^2 .$$

Thus for the negative exponential distribution with p.d.f.

$$f(x) = \mu e^{-\mu x} \qquad 0 \leqslant x < \infty \qquad (4.13)$$

we have

$$
\begin{aligned}
E(X) &= \mu \int_0^\infty x e^{-\mu x} \, dx \\
&= \mu \left[\frac{x e^{-\mu x}}{(-\mu)} - \frac{1 \cdot e^{-\mu x}}{(-\mu)^2} \right]_0^\infty \qquad \text{integrating by parts} \\
&= \frac{1}{\mu} .
\end{aligned}
$$

Also

$$
\begin{aligned}
E(X^2) &= \mu \int_0^\infty x^2 e^{-\mu x} \, dx \\
&= \mu \left[\frac{x^2 e^{-\mu x}}{(-\mu)} - \frac{2 x e^{-\mu x}}{(-\mu)^2} + \frac{2 e^{-\mu x}}{(-\mu)^3} \right]_0^\infty \qquad \begin{array}{l}\text{integrating by} \\ \text{parts twice}\end{array} \\
&= \mu \times \frac{2}{\mu^3} \\
&= \frac{2}{\mu^2} .
\end{aligned}
$$

Therefore

$$
\begin{aligned}
\sigma^2 &= E(X^2) - (E(X))^2 \\
&= \frac{2}{\mu^2} - \frac{1}{\mu^2} \\
&= \frac{1}{\mu^2} .
\end{aligned}
$$

APPENDIX 4.5 POISSON APPROXIMATION TO THE BINOMIAL DISTRIBUTION

Let $X \sim b(n, p)$ and let $n \to \infty$ and $p \to 0$ in such a way that $np = \lambda$. Then

$$
\begin{aligned}
\Pr(X = r) &= \frac{n!}{r!(n-r)!} \, q^{n-r} p^r \\
&= \frac{n(n-1) \ldots (n-r+1)}{r!} \left(1 - \frac{\lambda}{n} \right)^{n-r} \frac{\lambda^r}{n^r}
\end{aligned}
$$

$$= \frac{\lambda^r}{r!} \frac{n(n-1)\ldots(n-r+1)}{n^r} \left(1 - \frac{\lambda}{n}\right)^{-r} \times \left(1 - \frac{\lambda}{n}\right)^n$$

$$= \frac{\lambda^r}{r!} 1\left(1 - \frac{1}{n}\right)\ldots\left(1 - \frac{r-1}{n}\right) \times \left(1 - \frac{\lambda}{n}\right)^{-r} \times \left(1 - \frac{\lambda}{n}\right)^n.$$

As $n \to \infty$, the term $\left(1 - \frac{1}{n}\right)\ldots\left(1 - \frac{r-1}{n}\right) \times \left(1 - \frac{\lambda}{n}\right)^{-r}$ tends to 1 and

$\left(1 - \frac{\lambda}{n}\right)^n$ tends to $e^{-\lambda}$. Then

$$\Pr(X = r) = e^{-\lambda} \frac{\lambda^r}{r!} \quad \text{where } \lambda = np \tag{4.9}$$

APPENDIX 4.6   GENERATION OF NORMALLY DISTRIBUTED OBSERVATIONS

We generate n observations $U_1, U_2, \ldots, U_n$ randomly and independently uniformly distributed over the range 0 to 1.

Then

$$E(U_i) = \int_0^1 uf(u)du = \int_0^1 u.1.du = \frac{1}{2} \quad \text{since } f(u) = 1 \ \ 0 \leqslant u \leqslant 1$$
$$= 0 \ \text{ otherwise}$$

and

$$\sigma^2 = E(U_i^2) - (E(U_i))^2 = \int_0^1 u^2.1.du - \frac{1}{4} = \frac{1}{12}$$

Let $\quad \bar{U} = \frac{1}{n} \sum_{i=1}^n U_i$

Then $E(\bar{U}) = \frac{1}{2}$ and $\text{var}(\bar{U}) = \frac{1}{12n}$ (see equation (5.1)) and provided $n > 30$

$$\bar{U} \sim N\left(\frac{1}{2}; \frac{1}{12n}\right) \qquad \text{(see equation 5.2))}$$

and

$$Y = n\bar{U} = U_1 + U_2 + \ldots + U_n \sim N\left(\frac{n}{2}; \frac{n}{12}\right).$$

Thus

$$Z = \frac{Y - \dfrac{n}{2}}{\sqrt{\dfrac{n}{12}}} \sim N(0; 1)$$

and

$$X = \mu + \sigma Z = \mu + \frac{\sigma \left(Y - \dfrac{n}{2} \right)}{\sqrt{\dfrac{n}{12}}} \sim N(\mu; \sigma^2) \ .$$

In Program 4.3 the value $n = 40$ is used.

EXERCISES

Section A

1. The sex ratio of newborn animals of a certain breed is 100 female to 150 male. What is the probability that a random sample of 6 newborn animals will contain 2 males and 4 females? What is the most likely number of males in the sample?

2. At the turn of the century the probability of catching a certain tropical disease was 0.4. Using the binomial distribution calculate the probability that out of a party of ten men (a) none, (b) two or more, (c) eight or more, of them catch the disease. Discuss any assumptions made in calculating the probabilities in this way.

3. If the probability of a male birth is equal to that of a female birth, find the probability that, in a family with six children, the children are all of the same sex.

    There are ten families each with six children. Find the following probabilities.

    (i) The probability that no family (out of the ten) is such that its children are all of the same sex.
    (ii) The probability that all the families are such that the children in each family are of the same sex.

4. Six tomato plants were grown at each of 80 sites, and at each site the number attacked by spotted wilt disease was counted giving the following data:

Number of diseased plants (x)	0	1	2	3	4	5	6
Number of sites with x diseased plants	21	25	18	8	6	1	1

Fit a binomial distribution to the data.

5. The number of cornborer larvae on maize plants is known to follow a Poisson distribution with a mean of 0.5. Find the probability of finding (a) no, (b) more than one, (c) three or less, cornborer larvae on a plant selected at random.

6. In a series of experiments it is found that the probability that an animal will die as a result of being injected with a standard drug at a certain dose level is 0.02. If 100 animals are injected with this dose find the probability that at least three of them will die. Find the most likely number of animals to die in the experiment.

7. In a study of the nesting habitats of a certain species of bird the number of nests in certain areas of constant size were counted. The results are given in the following table.

No. of nests	0	1	2	3	4	5	6	Total
No. of areas	46	71	48	23	9	3	0	200

Calculate the mean and variance of this distribution and fit a Poisson distribution to the data.

8. When very accurate scientific instruments are being constructed it is often necessary that their assembly be carried out in 'dust-free' rooms. From one such 'dust-free' room 400 equal volumes of air were examined and the number of particles of dust in each volume was counted. The results are given in the following table.

No. of dust particles	0	1	2	3	4	5	6	Total
Frequency	88	138	98	50	16	8	2	400

Verify that the mean and variance of the distribution are approximately equal and fit a Poisson distribution to the data.

9. Find using tables of areas under the standard normal curve

    (i) $\Pr(-1.72 < Z < 0.84)$ (ii) $\Pr(-2.34 < Z < -1.88)$

    (iii) $\Pr(|Z| > 2)$ (iv) $\Pr(Z < 0.24)$

10. Find the values of z which satisfy the following

    (i) $\Pr(Z < z) = 0.5$ (ii) $\Pr(Z > z) = 0.05$

    (iii) $\Pr(|Z| > z) = 0.001$ (iv) $\Pr(Z < z) = 0.001$

11. The weights of capsules filled by a machine are normally distributed about a mean of 345.0 mg with a standard deviation of 7.0 mg. Let X mg be the weight of a capsule chosen at random from the output of the machine. Find

    (i) $\Pr(X < 360.0)$ (ii) $\Pr(330.0 < X < 370.0)$

    (iii) $\Pr(X > 365.0)$ (iv) $\Pr(X = 345)$

If a capsule chosen at random is found to have a weight of 375.0 mg what conclusion might you come to?

12. The weights of a certain type of rat used for experimental purposes are known to be normally distributed with a mean of 335.5 g and a standard deviation of 45.9 g. For a particular experiment it is required that the weight of each animal should be within the range 250.0 to 425.0 g. If 300 rats are chosen at random how many of them would you expect to be able to use for the experiment and how many of them would you expect to be rejected as overweight?

13. The weights of capsules are known to be normally distributed. Five per cent of the tablets have weights more than 366.5 mg while ten per cent have weights less than 337.2 mg. Find the mean and standard deviation of the distribution.

14. The amount of impurity in bottles of a preparation is normally distributed about a mean of 2.0% with a standard deviation of 0.3%. For a particular experiment the preparation is unsatisfactory if it contains more than 2.5% of impurity. If two bottles of the preparation are chosen at random find the probability that (a) they are both satisfactory, (b) one is satisfactory, (c) neither is satisfactory.

15. The weights of a batch of 100 capsules are reduced to a frequency distribution as follows

Class mid-value (mg)	337	342	347	352	357	362	367
Frequency	4	17	25	32	19	2	1

Calculate the mean and the standard deviation of the distribution.
Fit a normal distribution to the data.

16. The yields of penicillin contained in 200 batches of the drug were as given in the following table.

Yield (units/ml)	515–525	525–535	535–545	545–555	555–565
Frequency	10	50	90	30	20

Use probability paper

(i) to verify that the yields are normally distributed

(ii) to find the mean yield of the batches.

Then determine the standard deviation of the yield and estimate the percentage of batches whose penicillin concentration is less than 530 units/ml.

17. A market gardener is considering how many seeds of a certain plant he should sow. The plants will eventually have white or yellow blooms, the probability that a plant has a white bloom being 0.4 and the probability that it has a yellow bloom being 0.5. The remainder of the probability, 0.1, is to allow for plants which, for various reasons, e.g. wind and insect damage, do not flower. He wishes to be 80 per cent certain of obtaining 70 white blooms. How many seeds should he sow?

18. In a series of experiments it is found that the probability that an animal will die as a result of being injected with a standard drug at a certain dose level is 0.1.

(a) What is the probability that in a series of 10 experiments exactly 1 animal will die?

(b) What is the probability that in 100 such experiments at least 5 animals will die?

(c) What is the probability that the number of deaths in 100 experiments lies between 5 and 95 inclusive?

(In all cases assume a single experiment to mean the injection of 1 animal, and that all experiments are independent.)

19. The amount of impurity in batches of a drug is known to be approximately normally distributed about a mean of 1% with a standard deviation of ¼%. If the maximum allowable percentage impurity is 1½%, what proportion of batches will be unsatisfactory? Use the Poisson distribution to estimate the chance that of a week's production of 100 batches, 2 or more will be unsatisfactory.

20. The probability that the weight of an experimental animal exceeds 400 g is 0.2. If animals are drawn at random from a large population, find the probability that the first animal whose weight exceeds 400 g is selected on the third draw. Find the probability that at least three animals are selected before the first animal whose weight exceeds 400 g is drawn.

21. A weak long-lived radioactive source emits on average three rays per second. Find the probability that the interval between consecutive rays is (a) less than 0.5 second, (b) lies between 0.5 and 1.0 second. Find the probability that no ray is emitted in the second after the start of observation and that two rays are emitted in the following second.

Section B

1. The probability of a worker employed in a particular environment having a chronic cough is 0.75. If a sample of four workers is chosen at random from a large group, find the probability distribution of the number of workers, X, in the sample with chronic coughs. Find also the mean and variance of X.

2. In making electric fuses the failure rate is three per cent. Find, for a random sample of 10 fuses, the probabilities that (a) all are defective, (b) there are exactly two defectives, (c) there are at most two defectives.

3. The probability of a man of a certain age dying within ten years is 0.1. Out of a group of ten men of this age what is the probability that (a) none of them, (b) more than two of them, will die within ten years? What is the most likely number of the group to survive the period of ten years?

4. For a system to work satisfactorily for a particular period at least two out of four identical components in the system must not fail during the period. If the probability of a single component not failing during the period is 0.9999 find the probability that the system will fail.

5. The probability that a particular material contains more than a certain critical level of an impurity is 0.05. If five independent random samples of the material are chosen what is the probability that (a) all the samples are below the critical level, (b) all the samples are above the critical level, (c) at least one but not more than three of the samples are below the critical level?

6. In an industrial packaging process samples of five filled packages are taken at random intervals, weighed and the number of underfilled packages noted. During a test period the results of 1000 such samples are noted giving

No. of underfilled packages	0	1	2	3	4	5
Frequency of occurrence	777	198	23	2	0	0

Fit a binomial distribution to the data.

On average what percentage of packages are being underfilled during the test period?

7. The number of accidents of a certain type per month in a factory follows a Poisson distribution with a mean of 3. If X is the number of accidents per month, find $\Pr(X = r), r = 0, 1, 2, \ldots$.

8. The chance of a pedestrian in a certain age group being injured in a road accident is $1/1000$ in a year. Out of a group of 2000 such pedestrians find the probability that 3 or fewer will be injured within a year. What is the most likely number to be injured within a year?

9. The number of breakdowns in a continuously operating production line is Poisson with a mean number of 3.5 breakdowns per week (7 days). What is the probability of no breakdowns on a particular day? In how many days of the year (365 days) may two or more breakdowns be expected?

10. A weak long-lived radioactive source emits on average five rays per second. Find the most likely number of rays to be emitted per second by the source. What is the probability that two rays will be emitted in each of two consecutive seconds? What is the probability that four rays will be emitted in two seconds? Explain why the latter is greater than the former.

11. The following distribution of traffic is observed on a highway under free flow conditions.

Number of vehicles/ten-second period	0	1	2	3	4
Observed frequency	69	78	40	10	3

Fit a Poisson distribution to these data.

12. Find using tables of areas under the standard normal curve

(i) $\Pr(-1.82 < Z < 0.64)$ (ii) $\Pr(-1.96 < Z < -0.88)$
(iii) $\Pr(|Z| < 1)$ (iv) $\Pr(Z < 1.24)$

13. Find the values of z which satisfy the following

(i) $\Pr(Z > z) = 0.5$ (ii) $\Pr(Z < z) = 0.05$
(iii) $\Pr(|Z| > z) = 0.01$ (iv) $\Pr(Z > z) = 0.001$

14. The weights of capsules filled by a machine are normally distributed about a mean of 350.0 mg with a standard deviation of 7.0 mg. Let X mg be the weight of a capsule chosen at random from the output of the machine. Find

(i) $\Pr(X < 360.0)$ (ii) $\Pr(330.0 < X < 370.0)$ (iii) $\Pr(X > 365.0)$
(iv) $\Pr(X = 350)$

If a capsule chosen at random is found to have a weight of 375.0 mg, what conclusion might you come to?

15. The ozone concentration in a factory vent is normally distributed about a mean of 7.35 parts per hundred million with a standard deviation of 0.042. Find upper limits for the ozone concentration which in the long run will be exceeded (a) 5% of the time, (b) 1% of the time, and (c) 0.1% of the time.

16. The lengths of rods made by a machine are normally distributed. Five per cent of the rods have lengths of more than 102.07 cm and ten per cent have lengths less than 88.60 cm. Find the mean and standard deviation of the distribution.

17. The percentage of impurity in an alloy is normally distributed about a mean of 0.02% with a standard deviation of 0.003%. For a particular use the alloy is unsatisfactory if it contains more than 0.025% of the impurity. If two samples of the alloy are chosen at random find the probability that (a) they are both satisfactory, (b) one is satisfactory, (c) neither is satisfactory.

18. A mixture contains a mixture of white and black particles of the same size, the white particles constituting one per cent of the mixture. If a sample contains 500 particles what is the probability of finding fewer than three white particles in the sample?

19. Ten per cent of a certain type of crystal have flaws. Out of a random sample of 400 crystals find the probability that (a) at most 30, (b) between 30 and 50, (c) between 35 and 45, (d) 55 or more of the crystals have flaws.

20. The weights of 357 Breton Late Bronze Age socketed axes may be reduced to the frequency distribution tabulated below, where x is the class mid-value weight in grams.

x	252.5	257.5	262.5	267.5	272.5	277.5	282.5	287.5	292.5
f	2	6	12	20	32	35	35	64	51

x	297.5	302.5	307.5	312.5	317.5	322.5	327.5	332.5
f	35	13	13	13	11	5	5	5

Calculate the mean and standard deviation of the data and fit a normal distribution to the data.

Also test the fit of a normal distribution by using probability paper; estimate the mean and standard deviation of the distribution from your graph.

21. A large batch of screws contains ten per cent which are defective. Screws are drawn at random from the batch. Find the probability that (a) the first defective screw is selected on the second draw, (b) the first defective screw

is found between the third and fifth draws (inclusive). How many screws must be drawn for there to be a greater than 50 per cent chance of having at least one defective screw in the sample?

22. The number of customers arriving at a service station in a given interval of time has a Poisson distribution. The mean arrival rate is three in ten minutes. Starting from a given moment what is the probability that

    (a) the interval before the arrival of the first customer is nine minutes or more

    (b) there will be an interval of between three and five minutes before the arrival of the first customer

    (c) there will be no customers for three minutes and at least two customers in the following two minutes.

Section C†

1. Write a program to evaluate $\Pr(r \leqslant X \leqslant s)$ where $X \sim b(n, p)$ and $0 \leqslant r \leqslant s \leqslant n$.

2. Write a program to find the mean and standard deviation of a frequency distribution and to fit a binomial distribution with the same mean to the data.

3. Write a program to evaluate $\Pr(r \leqslant X \leqslant s)$ where $X \sim P(\lambda)$ and $0 \leqslant r \leqslant s$.

4. Write a program to find the mean and standard deviation of a frequency distribution and to fit a Poisson distribution with the same mean to the data.

5. Write a program to generate $\Pr(X = r)$ when (a) $X \sim b(n, p)$, (b) $X \sim P(\lambda)$ where $\lambda = np$. Use your program to test the Poisson approximation to the binomial distribution.

6. Using the appropriate subroutine from Program 4.3, or otherwise, write a program to find (a) $\Pr(X < x_1)$, (b) $\Pr(x_1 < X < x_2)$ (c) $\Pr(X > x_2)$ where $X \sim N(\mu; \sigma^2)$.

---

†Check the operation of the programs developed in this section on data from the examples of sections A and B, or worked examples in the text.

5

Elementary Statistical Inference

5.1 INTRODUCTION

In many applications of statistics we are interested in making *inferences* about *population* characteristics on the basis of observations made on a *random sample* of items from the population. The characteristics of interest may often be expressed in terms of population *parameters*, such as the population mean μ, or variance σ^2, or the proportion p of the population which has a certain characteristic. In other situations we may wish to make inferences about the difference between two (or more) populations, such as the difference between two population means μ_1 and μ_2.

The first question which arises is why we choose to make inferences based only on a sample taken from the population concerned. Of course if the characteristic or response concerned is the same for each member of the population then a sample of size one would suffice to determine its value for the population as a whole. However, in the application of statistical methodology we are concerned with responses or characteristics which vary for different members of the population under examination. In such situations there may be several reasons why it might be desirable to inspect a sample from the population rather than the whole population.

(a) The population, although finite, may be large enough to make one hundred per cent inspection too costly, or take too long a time.

(b) If the population is large, one hundred per cent inspection may not necessarily lead to a measurement of the population parameter which is absolutely precise, because of the risk of error in measurement or recording on repeating a routine measurement a large number of times.

(c) The measurement of the response may involve a destructive test, such as for example in the measurement of the tensile strength of a steel bar, or testing whether a match will strike or not.

(d) The whole population may not be available for inspection. For example a retrospective survey may be carried out on the basis of historical records which may be incomplete; in surveying a group of people, some of the group may be unavailable because of illness or travel.

(e) The population may be infinite. For example the response measured could be the result of an experiment which, for reasons outside the experimenter's control, may vary each time the experiment is repeated. The conceptual population here is the set of results which would be obtained on an infinite number of repetitions of the experiment, and is obviously not available to the experimenter.

Let us suppose that in a given situation it has been decided to make inferences concerning one or more population parameters on the basis of a sample taken from that population. It is important that in this case the sample is *random;* otherwise it would not be possible to make valid statistical inferences from observations or measurements on the sample. The problem of choosing a random sample from a finite population is considered in section 5.2. If the population being sampled is infinite we can design the experiment to take account of known sources of variation affecting the response under measurement, and by random allocation of treatments to the experimental units (e.g. variety of wheat and type of fertiliser to a plot of ground; particular dosage level of a type of drug to an experimental animal). However, the topic of design and analysis of experiments is outside the scope of this book. A useful introduction to the subject may be found in Stoodley, Lewis and Stainton (1980).

The following aspects of inference concerning one or more unknown population parameters will be discussed in this chapter

(a) obtaining an estimate of the parameter,
(b) obtaining some measure of precision of the estimate,
(c) carrying out statistical *hypothesis* (or *significance*) *tests* related to some predetermined value of the parameter.

The first requirement is some statistic which can be used as an *estimator* of the unknown parameter of the distribution. It is beyond the scope of this book to discuss the methods of choosing the 'best' estimator and we shall therefore assume without further investigation the statistic which we shall use as an estimator in a given situation. Such statistics, which are functions of the observations (but not, of course, of the unknown parameters which we are trying to estimate) will themselves be random variables. If we take a large number, N_0 say, of random samples, each of size n, and work out the statistic used as the estimator for each of the samples (this estimator could for example be the sample mean in the case of estimating a population mean), then N_0 values of the statistic will be obtained. These N_0 values will themselves form a distribution, referred to as the *sampling distribution* of the statistic. The properties of this sampling distribution may be inferred from those of the parent population. This in turn as we shall see allows us to make deductions about the unknown population parameter from the value of the estimator obtained from a *single* sample.

The following specific situations will be discussed:

(i) Sampling from a population with unknown mean μ where the sample size is large ($n \geqslant 30$). We wish to make inferences on μ (section 5.2). This situation will be used to illustrate many of the concepts of importance in statistical inference.

(ii) Sampling from two populations with unknown means μ_1 and μ_2 where the sample sizes n_1 and n_2 are large (n_1, $n_2 \geqslant 30$). We wish to make inferences on $\mu_1 - \mu_2$ (section 5.3).

(iii) Sampling from a population which can be divided into two classes A and B where the proportions p_A and p_B ($= 1 - p_A$) of A and B are unknown. We wish to make inferences on p_A from the information in a large sample (section 5.4).

(iv) Sampling from a normal population with unknown variance σ^2. We wish to make inferences on σ^2 (section 5.5).

(v) Sampling from two normal populations with unknown variances σ_1^2 and σ_2^2. Here we wish to make inferences on the ratio σ_1^2/σ_2^2 (section 5.6).

(vi) Sampling from a normal population with unknown mean μ (for all sample sizes). We wish to make inferences on μ (a) when the population variance is known (b) when the population variance is unknown (section 5.7).

(vii) Sampling from two normal populations with unknown means μ_1 and μ_2 and equal, unknown variances σ^2 (section 5.8). As in (ii) we wish to make inferences on the difference $\mu_1 - \mu_2$. The t-test for paired differences is also considered in this section.

In section 5.9 the problem of testing the goodness of fit of a distribution to a given set of data is discussed. Finally in section 5.10 the topic of Bayesian inference is briefly introduced.

As has already been mentioned the validity of the statistical inference methods to be discussed for the above situations depends on the principle of taking random samples, or carrying out appropriate randomisation in allocating treatments to experimental units, and also on the properties of the sampling distribution of the sample statistic used as the estimator. In particular the properties of the sampling distribution of the sample mean will be discussed in some detail in the next section.

5.2 INFERENCES CONCERNING THE POPULATION MEAN (LARGE SAMPLES)

5.2.1 Choosing a random sample and properties of the sampling distribution of the mean

Suppose that we have a batch of components and wish to estimate the mean lifetime μ weeks of the components in the batch on the basis of the lifetimes

$x_1, x_2, \ldots, x_n$ weeks of a random sample of size n drawn from the batch. It would seem reasonable to estimate μ by

$$\hat{\mu} = \bar{x}$$

where $\bar{x}$ is the sample mean and indeed this choice of estimate can be justified on theoretical grounds. However, often we should like to go further than obtaining this *point estimate* of μ; for example we might wish to give an indication of the precision of the estimate, or to test whether μ can be taken to equal some pre-determined value μ_0.

To undertake either of these tasks we must investigate the distribution of $\bar{X}$; this distribution is referred to as the *sampling distribution of the mean.*

To illustrate the concept of the sampling distribution of the mean consider the population of the lifetimes (in weeks) of 100 components tabulated in Table 5.1. We shall take this to be the population from which the samples are to be drawn. The population mean and standard deviation are $\mu = 53.92$ and $\sigma = 47.4423$.

Table 5.1

Lifetimes (to nearest week) of 100 components.

67	53	70	47	46	61	63	23	125	10
37	7	21	11	25	72	112	145	120	2
87	35	18	18	32	26	54	114	59	112
64	17	4	197	36	3	225	15	62	37
13	31	63	10	46	63	40	113	4	48
30	69	81	0	14	81	96	44	29	32
58	4	3	63	26	30	21	131	67	40
224	51	65	68	20	145	16	10	12	50
23	101	40	9	0	27	63	18	0	56
51	39	31	118	56	17	166	141	11	82

Reducing the population to a frequency distribution with classes of width 20 centred on $9.5, 29.5, \ldots$ leads to the distribution illustrated in Fig. 5.1.

Consider choosing (with replacement) a series of random samples each of size 4 from the population. Using a computer (or some electronic calculators) we can generate a random number r in the range 1 to 100; labelling the observations (by rows for example) from 1 to 100 we select the observation r and note its value and then replace it in the population, so that, if the random number r occurs again in the sequence of four random numbers, the observation can be selected again (in sampling without replacement an observation is included in a sample only on the first occasion its labelling number arises when the members of the sample concerned are being selected).

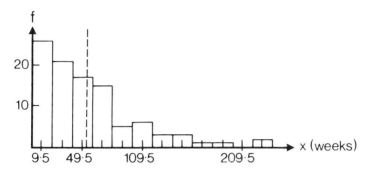

Fig. 5.1 — Distribution of the lifetimes of 100 components.

The random integers from 1 to 100 (inclusive) generated by the computer are such that in the long run the probability of occurrence of each integer is the same. The same result could be obtained (but with a lot more effort!) by placing pieces of paper labelled 1 to 100 in a hat and drawing them out (with or without replacement as required). The computer algorithm simulates this situation and random integers over any required range are easily obtained. The algorithm has the additional advantage that a given sequence may be repeated on setting suitable starting conditions for the algorithm.

Proceeding in this way we can take a large number N_0 of samples each of size $n = 4$ from the population. The resulting sample means $\bar{x}_1, \bar{x}_2, \ldots, \bar{x}_{N_0}$ will form a distribution called the *sampling distribution of the mean*. This distribution will have a mean and variance which we denote by $\mu_{\bar{X}}$ and $\sigma_{\bar{X}}^2$. It can be shown from theoretical considerations that

$$\mu_{\bar{X}} = \mu \text{ and } \sigma_{\bar{X}}^2 = \sigma^2/n \tag{5.1}$$

for all sample sizes n.

Using a computer, three simulation experiments of taking $N_0 = 1000$ samples from the population of Table 5.1 with $n = 4$, 16 and 36 were carried out. The means and standard deviations of the experimental distributions are tabulated in Table 5.2 where they are compared with the theoretical values. The sampling distributions of the means are illustrated graphically for the cases $n = 4$ and 36 in Figs. 5.2(a) and (b).

Table 5.2

Sample size n	Experimental values $\hat{\mu}_{\bar{X}}$	$\hat{\sigma}_{\bar{X}}$	Theoretical values $\mu_{\bar{X}}(=\mu)$	$\sigma_{\bar{X}} = \sigma/\sqrt{n}$
4	52.92	23.60	53.92	23.72
16	54.51	11.96	53.92	11.86
36	53.55	7.63	53.92	7.91

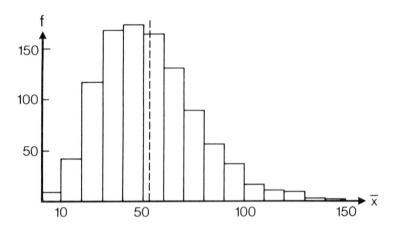

Fig. 5.2(a) – Sampling distribution of the means of 1000 samples of size 4 from the distribution of Fig. 5.1.

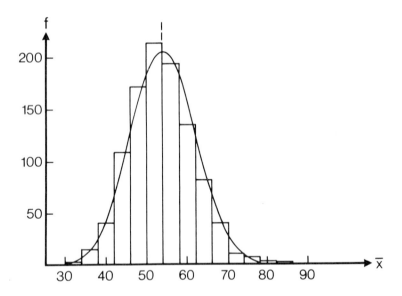

Fig. 5.2(b) – Sampling distribution of the means of 1000 samples of size 36 from the distribution of Fig. 5.1.

The continuous curve in Fig. 5.2(b) is that of a normal distribution with the same mean ($\mu = 53.92$) and variance ($\sigma^2/36 = 62.52$) as the sampling distribution of the mean. It will be seen that the normal distribution is a good fit to the sampling distribution of the mean for this value of n. This is an illustration of an important result which states that in general the sampling distribution of the mean is normal irrespective of the nature of the distribution of the population being sampled provided the sample size is large (in practice $n \geqslant 30$). Thus for $n \geqslant 30$ we have

$$\bar{X} \sim N(\mu; \sigma^2/n) \, . \tag{5.2}$$

For small sample sizes (in practice $n < 30$) the sampling distribution of the mean is normal only if the distribution of the population is itself normal, or nearly so.

Ex. 5.1. It is known that the mean weight of capsules filled by a machine is 350.0 mg with a standard deviation of 7.10 mg. Find the probability that a random sample of 36 capsules drawn from the output of the machine will have a mean weight of (a) between 348.0 and 352.0 mg, (b) greater than 352.5 mg.

Let $\bar{X}$ be the mean weight of the capsules in the sample.
Since $n = 36 > 30$ we have

$$\bar{X} \sim N(350.0; 7.1^2/36)$$

and

$$Z = \frac{\bar{X} - 350.0}{7.1/6} \sim N(0; 1) \, .$$

(a) $z_1 = \dfrac{348.0 - 350.0}{7.1/6} = -1.69, \qquad z_2 = \dfrac{352.0 - 350.0}{7.1/6} = 1.69 \, .$

$$\begin{aligned}
\text{Thus} \quad \Pr(348.0 < \bar{X} < 352.0) &= \Pr(-1.69 < Z < 1.69) \\
&= 2\Pr(0 < Z < 1.69) \\
&= 2 \times 0.4545 \\
&= 0.909 \, .
\end{aligned}$$

(b) $z = \dfrac{352.5 - 350.0}{7.1/6} = \dfrac{2.5 \times 6}{7.1} = 2.11 \, .$

Thus $\Pr(\bar{X} > 352.5) = \Pr(Z > 2.11) = 1.0 - 0.9826 \doteq 0.017.$

In the large sample case ($n \geqslant 30$) we may use the sampling distribution of the mean $\bar{X} \sim N(\mu; \sigma^2/n)$, to help us to set up *confidence intervals* for the unknown population mean μ and also to carry out *hypothesis tests* on the mean; these techniques will be described in the following sections.

5.2.2 Confidence intervals and limits

A *confidence interval* is used to give an idea of the *precision* of a point estimate. The principle of the technique is introduced in the next example.

Ex. 5.2. Machine filled capsules have an unknown mean weight μ mg with a standard deviation of 7.10 mg. Construct an interval, symmetrical about the mean weight $\bar{X}$ mg of a random sample of 36 capsules chosen from the output of the machine, such that the mean μ will lie in the interval 95% of the time for repeated samples of size 36.

Since $n = 36 > 30$ we have

$$\bar{X} \sim N\left(\mu; \frac{7.1^2}{36}\right) \text{ or } Z = \frac{\bar{X} - \mu}{7.1/6} \sim N(0; 1) .$$

With probability 0.95 we have

$$-z(0.025) < Z < z(0.025)$$

where $z(0.025)$ is the value of z such that the area in the upper tail of the distribution is 0.025. Thus, with probability 0.95

$$-1.96 < Z < 1.96$$

or

$$-1.96 < \frac{\bar{X} - \mu}{7.1/6} < 1.96 .$$

Re-arrangement of the limits (Appendix 5.1) leads to

$$\bar{X} - 1.96 \times \frac{7.1}{6} < \mu < \bar{X} + 1.96 \times \frac{7.1}{6} .$$

If we take a single sample of size 36 and calculate its mean $\bar{x}$, the limits $\bar{x} \pm 1.96 \times \dfrac{7.1}{6}$ are called 95% *confidence limits* for μ, and the interval

$\bar{x} - 1.96 \times \dfrac{7.1}{6}$ to $\bar{x} + 1.96 \times \dfrac{7.1}{6}$ is called a 95% *confidence interval* for μ.

In general if we take a sample of n ($\geqslant 30$) observations on a random variable X with population mean μ and variance σ^2, $100(1 - \alpha)\%$ confidence limits for μ are given by

$$\bar{x} \pm z(\alpha/2) \frac{\sigma}{\sqrt{n}} , \tag{5.3}$$

where $\bar{x}$ is the value of the mean for the sample concerned, and the confidence interval is

$$\bar{x} - z(\alpha/2)\,\frac{\sigma}{\sqrt{n}} \text{ to } \bar{x} + z(\alpha/2)\,\frac{\sigma}{\sqrt{n}}\;. \tag{5.4}$$

The *confidence coefficients* $z(\alpha/2)$ are obtained from Table A1(b). We note that as n increases the width of the confidence interval decreases, but if higher confidence is required the width increases.

If the population variance σ^2 is unknown it may be replaced in the expression for the confidence interval by its sample estimate $\hat{\sigma}^2$ where

$$\hat{\sigma}^2 = s^2 = \frac{1}{n-1}\sum_{i=1}^{n}(x_i - \bar{x})^2 = \frac{1}{n-1}\left\{\sum_{i=1}^{n}x_i^2 - \frac{\left(\sum\limits_{i=1}^{n}x_i\right)^2}{n}\right\}.$$

It was in order that the sample variance s^2 could be used without modification to estimate σ^2 that the divisor $(n-1)$ was used in the definition of s^2 (section 2.5).

Provided that the sample size is large enough the variability of $\hat{\sigma} = s$ may be neglected and replacing σ by s we may still assume

$$Z = \frac{\bar{X} - \mu}{s/\sqrt{n}} \sim N(0;1).$$

Hence in the case when σ is unknown the $100(1-\alpha)\%$ confidence interval is

$$\bar{x} - z(\alpha/2)\,\frac{s}{\sqrt{n}} \text{ to } \bar{x} + z(\alpha/2)\,\frac{s}{\sqrt{n}}\;. \tag{5.5}$$

Ex. 5.3. The heart rates of a sample of 40 dogs chosen at random from a particular population are measured and the sample mean is found to be 84 beats per minute with a standard deviation of 11 beats per minute. Estimate the mean heart rate for the population and find 99% confidence limits for the mean.

Let μ beats per minute be the mean heart rate for the population. Then

$$\hat{\mu} = \bar{x} = 84\;.$$

From (5.5) with $\alpha = 0.01$ the 99% confidence limits are

$$84 \pm z(0.005) \times \frac{11}{\sqrt{40}}$$

$$= 84 \pm 2.58 \times \frac{11}{\sqrt{40}}\;, \text{ using Table A1(b)}$$

$$= 84 \pm 4.5$$

$$= 79.5 \text{ and } 88.5\;.$$

5.2.3 Hypothesis tests on the population mean (large samples)

In many statistical inference problems we wish to test the hypothesis that the population mean μ for a random variable X is equal to some specified value μ_0. Here μ_0 may be a standard value for μ based on previous experience, a value derived from theoretical considerations, a value claimed by a manufacturer, or perhaps a value determined for a large control group. The hypothesis $H_0 : \mu = \mu_0$ is referred to as the *null hypothesis*.

Suppose that we have a sample of n ($\geqslant 30$) observations on X. Let $\bar{X}$ be the sample mean. Then as before

$$\bar{X} \sim N(\mu; \sigma^2/n) \ . \tag{5.2}$$

If H_0 is true, $\mu = \mu_0$ and

$$\bar{X} \sim N(\mu_0; \sigma^2/n)$$

and

$$Z = \frac{\bar{X} - \mu_0}{\sigma/\sqrt{n}} \sim N(0; 1) \ . \tag{5.6}$$

Z is the *test statistic* for the *hypothesis* (or *significance*) *test*.

The procedure followed at this stage depends upon the nature of the *alternative hypothesis* H_1 against which H_0 is being tested. We shall consider three forms for H_1 : (i) $H_1 : \mu > \mu_0$, (ii) $H_1 : \mu < \mu_0$, (iii) $H_1 : \mu \neq \mu_0$.

(i) Test of the hypothesis $H_0 : \mu = \mu_0$

        against $H_1 : \mu > \mu_0$.

In this case large positive values of z are evidence in favour of H_1. Values of z around zero are evidence in favour of H_0. Large negative values of z, while they are unlikely if H_0 is true, are even more unlikely if H_1 is true.

Let $\bar{x}$ be the value of $\bar{X}$ for the sample under consideration and let z be the corresponding value of Z.

We calculate $\alpha = \Pr(Z > z)$ using Table A1(a).

(a) If $\alpha > 0.05$ we say that z is *not significant;* in this situation we consider that there is no evidence for rejecting H_0 in favour of H_1.
(b) If $0.05 > \alpha > 0.01$ we say that z is *significant at the 5% level;* in this case we usually conclude that there is some evidence for rejecting H_0 in favour of H_1.
(c) If $0.01 > \alpha > 0.001$ we say that z is *significant at the 1% level;* this is generally interpreted as strong evidence for rejecting H_0 in favour of H_1.
(d) If $0.001 > \alpha$ we say that z is *significant at the 0.1% level;* in this case there is almost conclusive evidence for rejecting H_0 in favour of H_1.

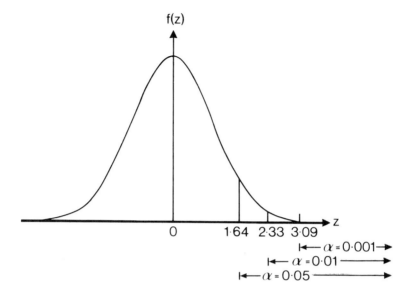

Fig. 5.3 – Critical regions for a one-tailed test (upper tail).

These regions are illustrated in Fig. 5.3, where the critical values of z have been obtained from Table A1(b).

This type of test is referred to as a *one-tailed* (upper tail) test.

(ii) Test of the hypothesis $H_0 : \mu = \mu_0$

against $H_1 : \mu < \mu_0$.

As in (i) we have $z = \dfrac{\bar{x} - \mu_0}{\sigma/\sqrt{n}}$. This time large negative values of z are evidence in favour of H_1 . We calculate

$$\alpha = \Pr(Z < z)$$

The values of α obtained are interpreted as in (i). The various regions of significance of z are illustrated in Fig. 5.4.

This test is a one-tailed (lower tail) test.

(iii)   Test of the hypothesis $H_0 : \mu = \mu_0$

against $H_1 : \mu \neq \mu_0$.

As in (i) and (ii) we calculate $z = \dfrac{\bar{x} - \mu_0}{\sigma/\sqrt{n}}$. This time both large positive and large negative values of z are evidence in favour of H_1 .

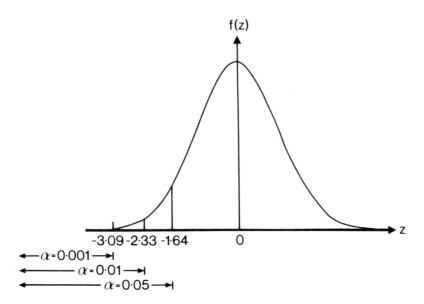

Fig. 5.4 – Critical regions for a one-tailed test (lower tail).

We calculate $\quad \alpha \;=\; \bar{\mathrm{P}}\mathrm{r}(|Z| > |z|)$

$\qquad\qquad\qquad = \mathrm{Pr}(Z < -|z|) + \mathrm{Pr}(Z > |z|)$

$\qquad\qquad\qquad = 2\mathrm{Pr}(Z > |z|) \quad \text{from symmetry.}$

Here '| |' means 'the numerical value of'.

Again the values of α obtained are interpreted as in (i).

The various ranges of α and the corresponding values of z are illustrated in Fig. 5.5.

This third type of test is referred to as a two-tailed test.

To find whether z is significant or not and, if it is, at what level, it is not necessary in practice to calculate the value of α. We simply compare the value of z for the experiment with the critical values illustrated in Figs. 5.3, 5.4 or 5.5. The critical values are obtained from Table A1(b).

As in the calculation of confidence limits, if σ is unknown it may be replaced by its sample estimate s without appreciably affecting the distribution of the test statistic; that is we may take

$$Z = \frac{\bar{X} - \mu_0}{s/\sqrt{n}} \sim N(0; 1) \tag{5.7}$$

if H_0 is true.

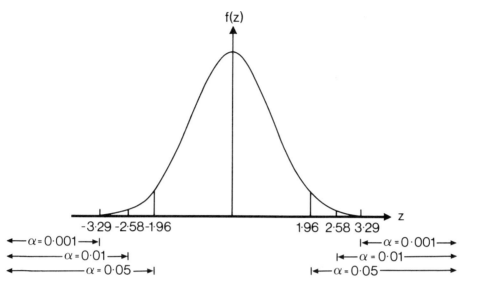

Fig. 5.5 – Critical regions for a two-tailed test.

Ex. 5.4 In an experiment involving a sample of 40 mice it is found that their mean loss in weight is 0.32 g with a standard deviation of 0.76 g. Is this change significantly different from zero?

Let $\overline{X}$ g be the mean change in weight for a sample of size 40 and μ g be the mean change in weight for the population.

Then we wish to test

$$H_0 : \mu = 0$$

against $H_1 : \mu \neq 0$.

A two-tailed test is used since the question posed does not specify that the change has to be in a specific direction (increase or decrease). Since the sample size is greater than 30 and σ^2 is unknown we have from (5.7) that, under H_0

$$Z = \frac{\overline{X} - \mu_0}{s/\sqrt{n}} \sim N(0; 1)$$

where $\mu_0 = 0$, $s = 0.76$ and $n = 40$. Then

$$z = \frac{-0.32 - 0}{0.76/\sqrt{40}}$$

$$= -2.66 \ .$$

Comparing with the critical values of z in Fig. 5.5 we have

$$-3.29 < z < -2.58$$

and z is significant at the 1% level.

Hence there is strong evidence in favour of H_1 that the experiment leads to a change in weight of the mice. The estimate of the change is

$$\hat{\mu} = \bar{x} = -0.32 \ .$$

If required, confidence limits could be calculated for μ using the given data. Note that although the change in weight is of statistical significance it is not necessarily of practical significance.

Ex. 5.5. A compound is known to contain 25% by weight of a certain element. The element is to be assayed by a method which is believed might over-estimate the amount of the element present. Fifty samples of the compound of equal weight are assayed using the method and the mean of the results is found to be 25.2% with a standard deviation of 1.0%. Are these results evidence in support of the supposed bias of the method?

Let $\mu\%$ be the population mean for the results obtained by the assay method and $\bar{X}\%$ the mean of a sample of 50 results. We wish to test

$$H_0 : \mu = 25$$

against $\quad H_1 : \mu > 25 \ .$

Here an upper one-tailed test is used since the question indicates that a bias, if it exists, will be positive.

From (5.7) the test statistic is

$$Z = \frac{\bar{X} - 25}{1/\sqrt{50}} \sim N(0; 1)$$

since $\mu_0 = 25$ and $n = 50$ here. Then

$$z = \frac{25.2 - 25.0}{1/\sqrt{50}} = 1.41 \ .$$

From Fig. 5.3 we see that $1.41 < 1.64$ and z is not significant. Hence there is no evidence that the method gives positively biased measurements of the amount of the element present in the compound.

The formulation of statistical hypothesis testing given above presents the results of the tests as evidence at various levels (none, some, strong, almost conclusive) in favour of the alternative hypothesis H_1.

An alternative formulation of statistical hypothesis testing is as follows. Before the experiment takes place we decide on a fixed value of α (usually, but not necessarily, one of the values 0.05, 0.01, 0.001). This will determine a critical region (see Figs 5.3, 5.4, 5.5, depending on whether the test is one-tailed (upper or lower) or two-tailed) in the tails of the distribution, such that if z falls within the critical region H_1 is accepted, otherwise H_0 is accepted. It will be noted that even if H_0 is true there is a probability α that Z will fall into the critical region. Thus α may be interpreted as the probability of accepting H_1 when H_0 is true; that is

$$\alpha = \Pr(\text{accepting } H_1 \mid H_0)$$

The probability of incorrectly accepting H_1 when H_0 is true could obviously be reduced by decreasing α. However there is a second type of error, that of accepting H_0 when H_1 is true, and reducing α increases the probability of this second type of error. In practice there will be costs associated with each type of wrong decision and the value of α used in any given situation will be determined by taking the relative values of these costs into account. If particular values of both errors are required the sample size may be increased until these values are attained. However, this could lead to an unacceptably large experiment.

This alternative method of formulation of hypothesis testing is useful in applications, such as quality control, where decisions have to be made as each sample is taken. In most scientific applications however the first approach to hypothesis testing considered is used, and it is this approach that will be followed in this book.

If the sample size n is large enough, any deviation from the null hypothesis will lead to a significant value for the test statistic. Thus in large experiments, results which are statistically significant may not correspond to changes of practical importance. On the other hand if the value of n is small, deviations from the null hypothesis which are of practical importance may not be detected because of the lack of sensitivity of the test. The order of magnitude of the changes which are detected and the sensitivity of the test may be judged by looking at the results of the test in conjunction with the appropriate point estimate and confidence limits for the parameter under investigation.

5.3 INFERENCES ON THE DIFFERENCE OF POPULATION MEANS (LARGE SAMPLES)

In this section we consider the situation where we have a random sample of size n_1 ($\geqslant 30$), with a mean of $\bar{x}_1$ and variance s_1^2, from a population P_1 with mean μ_1 and variance σ_1^2, and an independent random sample of size n_2 ($\geqslant 30$), with a mean of $\bar{x}_2$ and variance s_2^2, from a population P_2 with a mean μ_2 and variance σ_2^2.

Using the sample measurements we wish

(a) to estimate $\mu_2 - \mu_1$
(b) to find confidence limits for $\mu_2 - \mu_1$
(c) to test the null hypothesis $H_0 : \mu_2 = \mu_1$ against one of the alternatives:

$H_1 : \mu_2 \neq \mu_1, H_1 : \mu_2 > \mu_1$ or $H_1 : \mu_2 < \mu_1$.

The required estimate of $\mu_2 - \mu_1$ is $\bar{x}_2 - \bar{x}_1$. To proceed further we must obtain the sampling distribution of $\bar{X}_2 - \bar{X}_1$. It can be shown that, provided $n_1, n_2 \geqslant 30$, then

$$\bar{X}_2 - \bar{X}_1 \sim N \left(\mu_2 - \mu_1 \; ; \; \frac{\sigma_1^2}{n_1} + \frac{\sigma_2^2}{n_2} \right). \tag{5.8}$$

Using this distribution and normal distribution tables the required inferences may be made. The $100(1 - \alpha)\%$ confidence limits are then given by

$$\bar{x}_2 - \bar{x}_1 \pm z(\alpha/2) \sqrt{\left(\frac{\sigma_1^2}{n_1} + \frac{\sigma_2^2}{n_2} \right)}. \tag{5.9}$$

If σ_1^2 and σ_2^2 are unknown they may be replaced by their sample estimates $\hat{\sigma}_1^2 = s_1^2$ and $\hat{\sigma}_2^2 = s_2^2$ to give

$$\bar{x}_2 - \bar{x}_1 \pm z(\alpha/2) \sqrt{\left(\frac{s_1^2}{n_1} + \frac{s_2^2}{n_2} \right)}. \tag{5.10}$$

To test the null hypothesis $H_0 : \mu_2 = \mu_1$ we use the test statistic

$$Z = \frac{\bar{X}_2 - \bar{X}_1}{\sqrt{\left(\frac{\sigma_1^2}{n_1} + \frac{\sigma_2^2}{n_2} \right)}} \tag{5.11}$$

so that under H_0 $Z \sim N(0; 1)$. Here again σ_1^2 and σ_2^2 may be replaced by their sample estimates s_1^2 and s_2^2.

Ex. 5.6. A sample of 40 capsules filled by machine A have a mean weight of 350 mg with a variance of 50 mg^2 while a sample of 50 capsules filled by machine B have a mean weight of 340 mg with a variance of 45 mg^2.

(a) Estimate the difference between the mean weights $\mu_A - \mu_B$ of the capsules produced by the machines.
(b) Find 95% confidence limits for $\mu_A - \mu_B$.
(c) Test the null hypothesis $H_0 : \mu_A = \mu_B$ against the alternative hypothesis $H_1 : \mu_A \neq \mu_B$.

(a) The required estimate is

$$\widehat{\mu_A - \mu_B} = \bar{x}_A - \bar{x}_B$$
$$= 350 - 340$$
$$= 10$$

(b) From (5.10) the confidence limits are $\bar{x}_A - \bar{x}_B \pm z(0.025) \sqrt{\left(\dfrac{s_A^2}{n_A} + \dfrac{s_B^2}{n_B}\right)}$

$$= 350 - 340 \pm 1.96 \sqrt{\left(\frac{50}{40} + \frac{45}{50}\right)}$$

$$= 10 \pm 2.87$$

$$= 7.13 \text{ and } 12.87 .$$

(c) $H_0 : \mu_A = \mu_B$

$H_1 : \mu_A \neq \mu_B$

Under H_0

$$Z = \frac{\bar{X}_A - \bar{X}_B - 0}{\sqrt{\left(\dfrac{s_A^2}{n_A} + \dfrac{s_B^2}{n_B}\right)}} \sim N(0;1) \text{ from (5.11)} .$$

Hence $z = \dfrac{350 - 340}{\sqrt{\left(\dfrac{50}{40} + \dfrac{45}{50}\right)}}$

$$= \frac{10}{1.4663}$$

$$= 6.82 .$$

Since $z > 3.29$, it is significant at the 0.1% level and there is almost conclusive evidence in favour of $H_1 : \mu_A \neq \mu_B$. Hence if the machines are supposed to be producing capsules of the same weight, adjustment will be required to ensure that this condition is satisfied.

5.4 INFERENCES ON PROPORTIONS (LARGE SAMPLES)

Consider a large population, each item of which either possesses or does not possess an attribute A. Let p (unknown) be the proportion of items in the population which possess the attribute and q $(= 1 - p)$ be the proportion of those which do not. Let X be the number of items in a random sample of size n taken from the population which possess the attribute A. If n is very much smaller than the population size then, as we have seen in section 4.2, $X \sim b(n, p)$ and hence

$$\Pr(X = r) = \binom{n}{r} p^r q^{n-r}, \quad r = 0, 1, \ldots, n \quad . \tag{4.4}$$

Let $P = X/n$ be the proportion of items in the sample possessing the attribute A. Then it can be shown that P may be used as an estimator for p (that is $\hat{p} = P$).

If n is large enough (while still remaining very much smaller than the number of items in the population) the normal approximation to the binomial distribution may be used (section 4.7.2). Large n in this context means that $np \geqslant 5$ and $nq \geqslant 5$. In this case we have

$$X \sim N(np; npq) \qquad (5.12a)$$

which gives

$$P \sim N(p; pq/n) \ . \qquad (5.12b)$$

When p is unknown the variance may be estimated using the sample values P and $Q = 1 - P$ so that (5.12a) and (5.12b) lead to

$$X \sim N(np; nPQ) \qquad (5.13a)$$

and $$P \sim N(p; PQ/n) \qquad (5.13b)$$

where a continuity correction should be applied (section 4.7.2).

When the normal approximation is valid, significance tests on proportions may be carried out following the procedure of section 5.2.3. Approximate confidence limits for p can also be found using the distribution (5.13b).

Ex. 5.7. A machine is producing bolts of which a certain fraction are defective. The manufacturer claims that this fraction is 0.05. A random sample of 400 bolts is taken from a large batch and is found to contain 30 defective bolts. Does this indicate that the proportion of defectives is larger than that claimed by the manufacturer? Find 95% confidence limits for the proportion of defective bolts in the batch.

Let p be the proportion of defectives in the batch. Then

$$H_0 : p = 0.05$$
$$H_1 : p > 0.05 \ .$$

Let X be the number of defectives in a sample of size 400. Then under H_0

$$X \sim b(400, 0.05) \ .$$

Since $np = 400 \times 0.05 = 20 > 5$ and $nq = 400 \times 0.95 = 380 > 5$, the normal approximation to the binomial distribution is valid and we may take

$$X \sim N(20; 19)$$

with an appropriate continuity correction.

Following the argument of section 5.2.3 we calculate

$$\alpha = \Pr(X \geqslant 30)$$

$$= \Pr\left(Z > \frac{30 - \frac{1}{2} - 20}{\sqrt{19}}\right)$$

since the value $X = 30$ is to be included in the upper tail.

The last equation shows that the value of the test statistic should be taken as

$$z = \frac{30 - \frac{1}{2} - 20}{\sqrt{19}}$$

$$= 2.18 .$$

Since $z(0.05) = 1.64$ and $z(0.01) = 2.33$ the test statistic is significant at the 5% level and there is some evidence in favour of H_1 that the proportion of defectives in the batch is greater than 0.05.

Using (5.13b) the required confidence limits are

$$P \pm z(0.025) \sqrt{\frac{PQ}{n}}$$

$$= 0.075 \pm 1.96 \sqrt{\frac{0.075 \times 0.925}{400}}$$

$$= 0.049, 0.101 .$$

5.5 INFERENCES ON THE VARIANCE OF A NORMAL POPULATION

In this section we shall consider the problem of making inferences on the unknown variance σ^2 of a normal population, whose mean μ is also unknown, on the basis of a random sample of n observations $x_1, x_2, \ldots, x_n$ on the population. As might be expected the inferences are based on the sample variance

$$s^2 = \frac{1}{n-1} \sum_{i=1}^{n} (x_i - \bar{x})^2 . \tag{2.3}$$

The estimator $\hat{\sigma}^2$ of σ^2 is the sample variance s^2

$$\hat{\sigma}^2 = s^2 . \tag{5.14}$$

Further inferences on the variance are based on the sampling distribution of the statistic

$$U = (n-1) \frac{s^2}{\sigma^2} . \tag{5.15}$$

It can be shown from theoretical considerations that, providing the population is normal, the sampling distribution of U has a χ^2-*distribution* (read as 'chi-square'). The exact form of the p.d.f. of this distribution depends on the number ν of *degrees of freedom* (d.f.s) of the distribution, which here is worked out as $\nu = n - 1$. The term 'degrees of freedom' was borrowed from physics in the early days of development of statistics and remains a useful term in specifying parameters related to the sample size which are needed to completely determine a distribution. The p.d.f. of a chi-square distribution with four degrees of freedom (χ^2_4) is illustrated in Fig. 5.6. The lower and upper tail critical values of χ^2_ν may be found in Table A3.

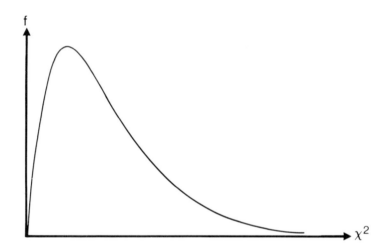

Fig. 5.6 – Probability density function for a chi-square distribution with $\nu = 4$.

Ex. 5.8. A random sample of ten components is chosen from a large batch of components produced by a machine and weighed. The sample variance is found to be 0.80 g^2. Find 95% confidence limits for the variance of the batch. Test whether the sample variance is significantly less than 1 g^2.

It has to be assumed for both sections of this example that the weights of the components are normally distributed.

The 95% confidence limits are obtained from the inequalities

$$\chi^2_9\,(0.975) < \frac{9s^2}{\sigma^2} < \chi^2_9\,(0.025) \ .$$

Here $\chi^2_9\,(0.975)$ is the value of chi-square with 9 degrees of freedom corresponding to an area of 0.025 in the lower tail.

From Table A3 we find χ_9^2 (0.975) = 2.70 and χ_9^2 (0.025) = 19.02, and substituting the sample value for s^2 we obtain

$$2.70 < \frac{7.2}{\sigma^2} < 19.02 \ .$$

Rearranging the limits (Appendix 5.2) leads to

$$0.38 < \sigma^2 < 2.67 \ .$$

Hence the required limits are 0.38 and 2.67.

For the second section of the example the hypotheses to be tested are

$$H_0 : \sigma^2 = 1$$

$$H_1 : \sigma^2 < 1 \ .$$

Under H_0 the test statistic $U = \dfrac{(n-1)s^2}{\sigma^2} = 9s^2 \sim \chi_9^2$ and the sample value of this statistic is

$$u = 9 \times 0.8 = 7.2 \ .$$

For a one-tailed (lower tail) test the critical value of χ_9^2 at the 5% level is χ_9^2 (0.95) = 3.33. Since the sample value of the test statistic is greater than this, the test statistic is not significant and there is no evidence in favour of the hypothesis $\sigma^2 < 1$.

5.6 INFERENCES ON THE RATIO OF VARIANCES OF TWO NORMAL POPULATIONS

The comparison of variances is particularly important in analysing the responses obtained from experimental designs (the technique of *analysis of variance*, see, for example, Stoodley, Lewis and Stainton (1980)). The technique is also used in the *analysis of regression* (section 6.2.1). The basic situation is that of taking a random sample of size n_1 from a normal population P_1 with unknown mean μ_1 and variance σ_1^2, and an independent random sample of size n_2 from a second normal population P_2 with unknown mean μ_2 and variance σ_2^2. As in the previous section we take

$$\hat{\sigma}_1^2 = s_1^2 \quad \text{and} \quad \hat{\sigma}_2^2 = s_2^2 \tag{5.16}$$

where s_1^2 and s_2^2 are the sample variances.

Assuming normal populations and independence, the sampling distribution of the ratio

$$F = \frac{s_1^2 / \sigma_1^2}{s_2^2 / \sigma_2^2} \tag{5.17}$$

may be obtained. The resulting distribution is *Fisher's F-distribution* F_{ν_1, ν_2}, where $\nu_1 = n_1 - 1$ and $\nu_2 = n_2 - 1$ are the degrees of freedom associated with the numerator and denominator of the F-ratio. The necessary critical values of F are tabulated in Table A4 for a range of values of ν_1 and ν_2.

It is not necessary to tabulate the lower tail critical values of F since it can be shown that

$$F_{\nu_1, \nu_2}(1 - \alpha) = \frac{1}{F_{\nu_2, \nu_1}(\alpha)} \, . \tag{5.18}$$

The p.d.f. of the F-distribution with $\nu_1 = 4$ and $\nu_2 = 6$ is illustrated in Fig. 5.7. For significance tests the null and alternative hypotheses can always be formulated in such a way that $s_1^2 > s_2^2$ so that $F > 1$. Then if the test statistic is significant, it will therefore fall into the upper tail of the distribution. Such a formulation is possible because either population may be selected as P_1, and F is a ratio of variances.

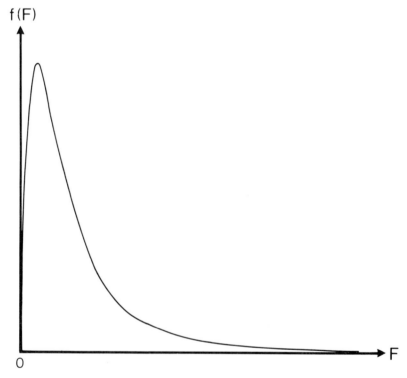

Fig. 5.7 – Probability density function of Fisher's F-distribution ($\nu_1 = 4, \nu_2 = 6$).

For a two-tailed test the critical value at the level α will be $F_{\nu_1, \nu_2}(\alpha/2)$ and, for a one-tailed test, $F_{\nu_1, \nu_2}(\alpha)$.

Ex. 5.9. The weights (in kg) of random samples of ten packages from each of two large batches A and B are tabulated below.

Batch A	50.8	51.0	49.5	52.1	51.8	47.4	51.5	48.2	49.0	48.0
Batch B	49.3	48.9	49.2	50.0	48.8	49.5	49.2	49.6	48.8	47.5

Is there evidence of a difference in the variance of the weights of the packages in the two batches?

Let $x_1, \ldots, x_{10}$ kg be the weights of the packages from batch A and let $y_1, \ldots, y_{10}$ kg be the weights of the packages from batch B.

Let σ_A^2 kg^2 and σ_B^2 kg^2 be the variances of the weights in the two batches. We have to assume that the distribution of the weights in both batches is normal.

From the given data $\sum_{i=1}^{10} x_i = 499.3$, $\sum_{i=1}^{10} x_i^2 = 24956.79$.

Hence

$$s_A^2 = \frac{1}{9}\left(24956.79 - \frac{499.3^2}{10}\right) = 2.9712$$

and

$$\sum_{i=1}^{10} y_i = 490.8, \quad \sum_{i=1}^{10} y_i^2 = 24092.52 .$$

Hence

$$s_B^2 = \frac{1}{9}\left(24092.52 - \frac{490.8^2}{10}\right) = 0.4507 .$$

Since $s_A^2 > s_B^2$ we formulate the null and alternative hypotheses as

$$H_0 : \sigma_A^2 = \sigma_B^2$$
$$H_1 : \sigma_A^2 \neq \sigma_B^2 .$$

Then under H_0 $F = \dfrac{s_A^2}{s_B^2}$ will have Fisher's F-distribution with $\nu_1 = \nu_2 = 9$.

The sample value of $F = \dfrac{2.9712}{0.4507} = 6.59.$

From Table A4 $F_{9,9}(0.025) = 4.03$, $F_{9,9}(0.005) = 6.54$.

Hence there is strong evidence in favour of $H_1 : \sigma_A^2 \neq \sigma_B^2$.

Ex. 5.10. It is known that over a long period the weights of aspirin tablets produced by a machine have been normally distributed with a standard deviation of 3.16 mg. A random sample of ten tablets taken from a large batch produced by the machine has a standard deviation of 2.56 mg. Is there evidence that the variance of the weights of the tablets produced by the machine is less than usual?

Let σ^2 mg^2 be the variance of the tablets in the batch. We wish to test

$$H_0 : \sigma^2 = 9.9856$$

$$H_1 : \sigma^2 < 9.9856$$

There are two ways of carrying out this test

(a) by using the chi-square test

$$U = \frac{9s^2}{\sigma^2} \sim \chi_9^2 .$$

The sample value of the test statistic is

$$u = 9 \times \frac{2.56^2}{3.16^2} = 5.91.$$

For a lower tailed-test $\chi_9^2 (0.05) = 3.33$.

As the value of u is greater than this, there is no evidence that the variance of the batch is less than the usual value.

(b) by using an F-test

$$F = \frac{s^2}{\sigma^2} \text{ with } \nu_1 = 10 - 1 = 9 \text{ and } \nu_2 = \infty.$$

Here ν_2 is taken to be infinity since the given value of σ^2 is based on a very large number of observations. The sample value of F is

$$F = \frac{2.56^2}{3.16^2} = 0.66 .$$

The critical value for a lower tailed test at the 5% level is

$$F_{9, \infty} (0.95) = \frac{1}{F_{\infty,9}(0.05)} = \frac{1}{2.71} = 0.37, \text{ using (5.18)} .$$

Since the sample value of the test statistic (0.66) is greater than the critical value at the 5% level we once again conclude that there is no evidence in favour of H_1.

Note that the tests are completely equivalent, since the sample values of the test statistics and the critical values against which they are tested are both in the ratio 9 to 1.

5.7 INFERENCES ON THE POPULATION MEAN (SMALL SAMPLES)

5.7.1 Student's t-distribution

For small sample sizes (in practice $n < 30$) the result that the sampling distribution of the mean is normal will only hold if the population from which the sample is drawn is itself normal, or nearly so. We shall therefore assume in this section, concerned with small samples, that the parent population is normal (or nearly normal) so that (as in section 5.2)

$$\bar{X} \sim N(\mu; \sigma^2/n) \ . \tag{5.2}$$

The point estimate $\hat{\mu}$ of μ is still taken as the sample mean $\bar{x}$. Also we have

$$Z = \frac{\bar{X} - \mu}{\sigma/\sqrt{n}} \sim N(0; 1) \qquad \text{(from 5.2).}$$

If σ is known, confidence limits for μ may be calculated and hypothesis tests on μ carried out using the above result as in the large sample case (sections 5.2.2 and 5.2.3).

However, if σ is unknown we can no longer assume that the statistic $\dfrac{\bar{X} - \mu}{s/\sqrt{n}}$ will have a standard normal distribution, since the estimator s of σ is subject to appreciable variability from sample to sample when the sample size is small. However, with the assumption of a normal population it is possible to deduce the sampling distribution of the statistic

$$t = \frac{\bar{X} - \mu}{s/\sqrt{n}} \ . \tag{5.19}$$

This distribution is referred to as *Student's t-distribution*. The exact form of the p.d.f. of the distribution depends on the sample size n through $\nu = n - 1$; ν is the number of degrees of freedom for the t-distribution.

The p.d.f. curve of the t-distribution with $\nu = 4$ is illustrated in Fig. 5.8. It is symmetrical about the axis $t = 0$. For small ν (n) more area is in the tails of the distribution, but as ν increases the tails of the distribution are pulled in and the curve approaches that of the p.d.f. of the standard normal distribution. For practical purposes the t-distribution may be approximated by the normal distribution if n is greater than or equal to 30.

5.7.2 Confidence limits for the mean (small samples)

Let $\bar{x}$ be the mean of a sample of n observations on a normally distributed random variable X with unknown mean μ and unknown variance σ^2. Then confidence limits for μ may be established as in section 5.2.2, but using critical

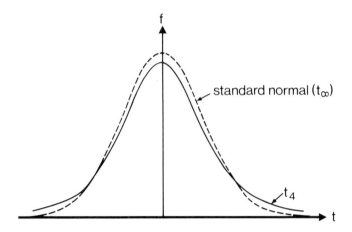

Fig. 5.8 – Probability density functions of Student's t-distribution with $\nu = 4$ and the standard normal distribution.

values taken from Table A2 for the t-distribution instead of from normal distribution tables. The $100(1 - \alpha)\%$ confidence limits are then given by

$$\bar{x} \pm t_\nu(\alpha/2) \, \frac{s}{\sqrt{n}} \, . \tag{5.20}$$

Ex. 5.11. The stroke volume X ml per heart-beat in a particular breed of dog is approximately normally distributed with a mean of μ ml and a variance of σ^2 ml^2. Measurement of this volume for eight dogs are

$$23 \quad 40 \quad 44 \quad 39 \quad 38 \quad 40 \quad 48 \quad 36$$

Find 95% confidence limits for μ.

From the data $\bar{x} = 38.5$ and $s = 7.2899$.
From (5.20) the 95% confidence limits for μ are

$$\bar{x} \pm t_{n-1}\,(0.025) \, \frac{s}{\sqrt{n}}$$

$$= 38.5 \pm 2.365 \times \frac{7.2899}{\sqrt{8}}$$

$$= 32.4 \text{ and } 44.6.$$

5.7.3 Hypothesis tests on the mean (small samples)
Under $H_0 : \mu = \mu_0$ the test statistic

$$t = \frac{\bar{X} - \mu_0}{s/\sqrt{n}} \tag{5.19}$$

will have Student's t-distribution with $\nu = n - 1$ d.f.s. Using this statistic the hypothesis H_0 may be tested against H_1, where H_1 is one of $\mu > \mu_0, \mu < \mu_0$ or $\mu \neq \mu_0$, using the method described in section 5.2.3. However, because only certain critical values of t are tabulated in Table A2, we must use the method of comparing the sample value for t with the appropriate critical values to find whether the value of t is not significant, significant at the 5% level and so on.

Ex. 5.12. The mean weight of a random sample of twenty tablets is found to be 0.33086 g with a standard deviation of 3.16×10^{-3} g. Test the hypothesis that the sample has come from a population with a mean weight of 0.33142 g

(a) if the population standard deviation is unknown
(b) if the population standard deviation is known to be 3.14×10^{-3} g.

In either case we must assume that the population is at least approximately normal. Let μ g be the population mean. Then we wish to test

$$H_0 : \mu = 0.33142$$

against $$H_1 : \mu \neq 0.33142 \ .$$

The form of H_1 tells us that a two-tailed test is required (section 5.2.3).

(a) The test statistic is

$$t = \frac{\bar{X} - 0.33142}{\dfrac{s}{\sqrt{20}}}$$

which under H_0 will have a t-distribution with 19 d.f.

The sample value for t is

$$\frac{0.33086 - 0.33142}{\dfrac{3.16 \times 10^{-3}}{\sqrt{20}}}$$

$$= - 0.79 \ .$$

$t_{19}(0.025) = 2.093$ from Table A2.
    Since $-2.093 < -0.79 < 2.093$ the result is not significant and there is no evidence that the population mean differs from 0.33142 g.

(b) Since the population standard deviation is known ($\sigma = 3.14 \times 10^{-3}$) we can use the test statistic

$$Z = \frac{\bar{X} - 0.33142}{\dfrac{3.14 \times 10^{-3}}{\sqrt{20}}}$$

which if H_0 is true will have a standard normal distribution.

The sample value for z is $\dfrac{0.33086 - 0.33142}{3.14 \times 10^{-3}/\sqrt{20}}$

$$= -0.80 \ .$$

Now $z(0.025) = 1.96$. Since $-1.96 < -0.80 < 1.96$ the result is not significant and again there is no evidence that the population mean differs from 0.33142 g. Note that we are making the additional assumption here that, even if the mean has changed, the standard deviation has not.

It is preferable to use a test based on the normal distribution where we can, as this is equivalent to a t-test on '∞' d.f.s; in general the more d.f.s (or the larger the sample size) that are available the less will be the risk of making incorrect inferences from the observations.

Ex. 5.13. Glucose treatment is expected to lengthen the sleep time of mice. The mean hexobarbital sleep time of mice without treatment is known to be approximately normally distributed about a mean of 28.5 minutes. Measurements of the corresponding sleep time of ten treated mice are (in minutes)

$$43 \quad 65 \quad 53 \quad 26 \quad 47 \quad 49 \quad 50 \quad 57 \quad 44 \quad 57$$

Is the mean of the sample significantly greater than 28.5 minutes?

From the above data $\bar{x} = 49.1$ and $s = 10.5140$

Let μ min be the (population) mean sleep time for the treated mice. Then

$$H_0 : \mu = 28.5$$

$$H_1 : \mu > 28.5$$

The form of H_1 implies that a one-tailed (upper tail) test is to be used.

Under H_0 $\qquad\qquad\qquad t = \dfrac{\bar{X} - 28.5}{\dfrac{s}{\sqrt{10}}}$

has Student's t-distribution with 9 d.f.

The sample value for t is $\qquad t = \dfrac{49.1 - 28.5}{10.5140/\sqrt{10}}$

$$= \quad 6.20 \ .$$

From Table A2 $\quad t_9(0.05) = 1.833, t_9(0.01) = 2.821, t_9(0.001) = 4.297$.

Since $6.20 > 4.297$ the result is significant at the 0.1% level and there is almost conclusive evidence for the hypothesis that the glucose treatment lengthens the sleep time.

5.8 INFERENCES ON THE DIFFERENCE OF POPULATION MEANS (SMALL SAMPLES)

5.8.1 Independent samples from two populations

In experimental work a problem frequently encountered is that of making inferences on the difference of the means of two populations P_1 and P_2 on the basis of small samples taken from the populations, when the population variances are also unknown. Two experiments in which this situation arises are briefly described below.

The first experiment concerns the effect of glucose treatment on the sleep time of mice; first we choose a sample of, say, ten mice at random and after giving them glucose for a prescribed period, measure their hexobarbital sleep time. Then we select a further random sample of mice, often, but not necessarily, with the same number of mice as the first. These are not given glucose treatment, but otherwise treated in the same way as the first sample, and their hexobarbital sleep time is also measured. On the basis of the measured sleep times of the two samples we wish to test the hypothesis that the glucose treatment extends the sleep time.

The second experiment is designed to test the effect of fenfluramine on the noradrenalin (NA) content of mouse brain. The NA contents of the brains of an untreated random sample of mice are first measured. Then the NA contents of the brains of a second random sample of mice treated with a specified dosage of fenfluramine (in mg/kg body weight) are measured. On the basis of these measurements we test the hypothesis that fenfluramine affects the brain NA content.

The situation in such experiments may be described as follows.

Let $\bar{x}_1$ be the mean of a sample of n_1 observations from a population P_1 with unknown mean μ_1 and unknown variance σ_1^2 and let $\bar{x}_2$ be the mean of an independent sample of n_2 observations from a population P_2 with unknown mean μ_2 and unknown variance σ_2^2.

Then a point estimate of $\mu_2 - \mu_1$ may be obtained from the difference of the sample means $\bar{x}_2 - \bar{x}_1$. To proceed further we must make two additional assumptions

(a) the populations P_1 and P_2 are normally distributed (or nearly so)
(b) $\sigma_1^2 = \sigma_2^2 = \sigma^2$, say, where the common value σ^2 is unknown.

Under these conditions it can be shown that the statistic

$$ t = \frac{\bar{X}_2 - \bar{X}_1 - (\mu_2 - \mu_1)}{\hat{\sigma}\sqrt{\left(\dfrac{1}{n_1} + \dfrac{1}{n_2}\right)}} \tag{5.21}$$

has Student's t-distribution with $\nu = n_1 + n_2 - 2$ d.f.s.

Here

$$\hat{\sigma}^2 = \frac{(n_1 - 1) s_2^2 + (n_2 - 1) s_2^2}{n_1 + n_2 - 2} \tag{5.22}$$

where s_1^2 and s_2^2 are the sample variances, is an estimate of the common variance σ^2 based on the information in both samples. The number of degrees of freedom $\nu = n_1 + n_2 - 2$ is the sum of the degrees of freedom $\nu_1 = n_1 - 1$ and $\nu_2 = n_2 - 1$ associated with each sample.

Ex. 5.14. It is believed that glucose treatment will extend the sleep time of mice. In an experiment to test this hypothesis ten mice selected at random are given glucose treatment and are found to have a mean hexobarbital sleep time of 47.2 min with a standard deviation of 9.3 min. A further sample of ten untreated mice are found to have a mean hexobarbital sleep time of 28.5 min with a standard deviation of 7.2 min. Are these results significant evidence in favour of the hypothesis?

Find 95% confidence limits for the population mean difference in sleep time. State any assumptions made concerning the data in carrying out the test and finding the limits.

Let μ_1, $\bar{x}_1$ min be the population and sample mean sleep times for the untreated mice. Let μ_2, $\bar{x}_2$ min be the population and sample mean sleep times for the treated mice.

We wish to test $H_0 : \mu_2 = \mu_1$

against $H_1 : \mu_2 > \mu_1$

From the form of H_1 a one-tailed (upper tail) test is required. Under H_0

$$t = \frac{\bar{X}_2 - \bar{X}_1 - 0}{\hat{\sigma} \sqrt{\frac{1}{n_1} + \frac{1}{n_2}}} \sim \text{Student's } t \text{ with } n_1 + n_2 - 2 \text{ d.f.s}$$

(since $\mu_2 - \mu_1 = 0$ under H_0).

$$\hat{\sigma}^2 = \frac{(n_1 - 1) s_1^2 + (n_2 - 1) s_2^2}{n_1 + n_2 - 2} = \frac{9 \times 9.3^2 + 9 \times 7.2^2}{18} = 69.165,$$

giving

$$t = \frac{47.2 - 28.5}{\sqrt{69.165 \left(\frac{1}{10} + \frac{1}{10}\right)}} = 5.03 \text{ with } 18 \text{ d.f.}$$

From Table A2, $t_{18}(0.05) = 1.734$, $t_{18}(0.01) = 2.552$, $t_{18}(0.001) = 3.610$.

Hence the result is significant at the 0.1% level and there is almost conclusive evidence for accepting the hypothesis that the treatment increases the sleep time.
Ninety five per cent confidence limits for $\mu_2 - \mu_1$ are given by

$$\bar{x}_2 - \bar{x}_1 \pm t_\nu(0.025)\hat{\sigma} \sqrt{\left(\frac{1}{n_1} + \frac{1}{n_2}\right)} \qquad (5.23)$$

which are derived from re-arranging the inequalities

$$-t_\nu(0.025) < \frac{\bar{X}_2 - \bar{X}_1 - (\mu_2 - \mu_1)}{\hat{\sigma}\sqrt{\left(\frac{1}{n_1} + \frac{1}{n_2}\right)}} < t_\nu(0.025)$$

and replacing $\bar{X}_1$, $\bar{X}_2$ and $\hat{\sigma}$ by their values calculated from the sample observations (cf. section 5.2.2).
Hence the required limits are

$$47.2 - 28.5 \pm 2.101 \sqrt{69.165\left(\frac{1}{10} + \frac{1}{10}\right)}$$

$$= \quad 18.7 \pm 7.8$$

$$= \quad 10.9, 26.5 \ .$$

In carrying out the above calculations we have assumed that the sleep times for both the population of treated and the population of untreated mice are normally distributed with equal variances for the two populations.

If necessary both these assumptions can be tested statistically.

To test the assumption of normality, more observations should be taken. Then one of the methods of section 4.6.1 may be applied. To test the assumption of equality of variance, the F-test introduced in section 5.6 can be applied. If either or both of the assumptions are untrue, alternative test procedures are available (see for example Siegel (1956) or Stoodley et al. (1980)).

5.8.2 Matched pairs

In experiments to study the response of experimental units to a treatment it is often possible to eliminate, or at least reduce, the effect of extraneous factors relating to the experimental units (such as for example age, sex, medical history) by dividing the units into pairs which are matched as far as possible with regard to these factors. One member of each matched pair is then chosen at random and receives the treatment under investigation and the other member is used as a control. In some situations it is possible to use a unit as its own control provided the treatment does not have any prolonged or lasting after-effects. Identical twins make good matched pairs for most experiments of this type. The difference between the responses of the two members of the pair then gives a measure of the effectiveness of the treatment. The method of analysis is illustrated in the example below.

Ex. 5.15. Six dogs were treated with specified dosages of aldosterone and spironolactone and, under carefully controlled conditions, their urine output (in ml/kg/5 h) was measured over a period of five hours following administration of the drugs. The experiment was repeated one week later on the same six dogs without use of the drugs. The measurements obtained in the two experiments are tabulated below

Dog	1	2	3	4	5	6
Without drugs	12.0	8.3	11.1	11.5	10.0	12.8
With drugs	7.3	12.8	13.1	11.1	19.4	9.0

Test whether the use of drugs affects the volume of urine output.

Let y_{ij} be the response from dog j undergoing treatment $i (i = 1$ without drugs, $i = 2$ with drugs). Then we may model the responses by

$$y_{ij} = \mu_i + \delta_j + \epsilon_{ij}, \quad i = 1, 2; j = 1, 2, \ldots, n \; (= 6 \text{ here})$$

where μ_i is the effect due to treatment i, δ_j the effect due to dog j and ϵ_{ij} represents random experimental error.

Let $\quad d_j = y_{2j} - y_{1j}$

$$= \mu_2 + \delta_j + \epsilon_{2j} - (\mu_1 + \delta_j + \epsilon_{1j})$$

$$= \mu_2 - \mu_1 + \epsilon_j$$

where $\epsilon_j = \epsilon_{2j} - \epsilon_{1j}$.

By differencing we have eliminated the unwanted 'dog effects' δ_j and the differences d_j depend only on the difference $\mu_2 - \mu_1$ between treatments and the random experimental error ϵ_j. Let $\bar{d}$ and s_d be the mean and standard deviation of these differences. Then if the ϵ_js are identically and independently normally distributed (or nearly so), with mean zero and (unknown) variance σ^2 it can be shown that the statistic

$$t = \frac{\bar{d} - (\mu_2 - \mu_1)}{\dfrac{s_d}{\sqrt{n}}} \tag{5.24}$$

has Student's t-distribution with $\nu = n - 1$ d.f.s, where n is the number of matched pairs in the experiment ($= 6$ in this example); we are assuming here that each set of measurements may be regarded as a sample from a normal population. This statistic may be used to set up confidence limits for $\mu_2 - \mu_1$

and also to test the hypothesis $H_0 : \mu_2 - \mu_1 = 0$, that there is no treatment effect. In this example we wish to test

$$H_0 : \mu_2 = \mu_1$$

against $H_1 : \mu_2 \neq \mu_1$

From the form of H_1 we see that a two-tailed test is required.

Since the alternative hypothesis is $H_1 : \mu_2 \neq \mu_1$ it does not matter whether we define d_j as $y_{2j} - y_{1j}$ or $y_{1j} - y_{2j}$; we take the responses without drugs from those with drugs to give

Dog	1	2	3	4	5	6
d_j	−4.7	4.5	2.0	−0.4	9.4	−3.8

Then

$$\sum_{j=1}^{6} d_j = 7 \qquad \text{and} \quad \bar{d} = 1.1\dot{6},$$

$$\sum_{j=1}^{6} d_j^2 = 149.3 \quad \text{and} \quad s_d = 5.313.$$

Under H_0 $\qquad t = \dfrac{\bar{d} - 0}{\dfrac{s_d}{\sqrt{6}}} \sim$ Student's t with 5 d.f.

$$t = \frac{1.1\dot{6}}{5.313/\sqrt{6}} = 0.54 \ .$$

Since $t_5(0.025) = 2.571$ the result is not significant and there is no evidence that the drug treatment has an effect on the volume of urine output.

Some care must be taken to distinguish between situations where the t-test for difference of means is appropriate and where the t-test for matched pairs is appropriate.

5.9 TESTING FOR GOODNESS OF FIT

In example 4.4 (fitting a binomial distribution), 4.9 (fitting a Poisson distribution), and 4.15 (fitting a normal distribution) visual comparison of the columns of observed and expected frequencies showed that there was close agreement between them. A statistical test of the *goodness of fit* may be carried out as follows. Let k be the number of classes over which the expected (e_i) and observed

(o_i) frequencies are being compared. The χ^2 (chi-square) statistic is then calculated from

$$\chi^2 = \sum_{i=1}^{k} \frac{(e_i - o_i)^2}{e_i} \; . \tag{5.25}$$

Provided each of the expected frequencies e_i is greater than or equal to 5 and the fitted distribution is the correct one, this statistic will (to a good approximation) have a χ^2-distribution. As we have already seen (section 5.5) the exact form of the p.d.f. of this distribution depends on the number of degrees of freedom, ν, associated with the problem under consideration. The method of determining ν for the application of the χ^2-distribution to the testing of goodness of fit of the observed and expected distributions is given below.

Note that if all the observed and expected frequencies coincided exactly, the value of χ^2 would be zero, while large (positive or negative) differences between the observed and expected frequencies for any of the classes will give large positive values of χ^2.

Hence small values of χ^2 indicate a good fit of the observed and expected distributions and large values of χ^2 indicate a poor fit.

The critical values $\chi^2_\nu(\alpha)$ of χ^2 may be found from Table A3 for a range of values of ν, the degrees of freedom, and several values of α. If we take for example, $\alpha = 0.05$ and the correct distribution is being fitted to the observed data, the calculated value of χ^2 will exceed the critical value $\chi^2_\nu(0.05)$ only 1 in 20 times by chance. Hence, if for a given set of observed data the value of χ^2 calculated from the observed frequencies and the expected frequencies obtained from fitting a particular distribution is greater than $\chi^2_\nu(0.05)$, we prefer to conclude that the fit of the observed and expected distributions is not satisfactory.

In order to satisfy the condition that each expected frequency is greater than 5, classes with expected frequencies of less than 5 must be combined with adjacent classes until this condition is met. The observed frequencies in the' corresponding classes must also be combined. Thus in example 4.4, no combination of classes is necessary since each expected frequency is greater than 5; in example 4.9, the classes $X = 5$ and $X \geqslant 6$ must be combined to give an expected frequency of 5.3 corresponding to an observed frequency of 5; and in example 4.15, the expected frequencies in the two tail classes must be combined with those in the adjacent classes to give expected frequencies of 10.8 and 9.9 corresponding to observed frequencies of 10 in each case.

The number of degrees of freedom ν are then calculated as follows:

$$\nu = k - 1 - m \tag{5.26}$$

Here k is the number of classes remaining after any necessary combination of classes has been carried out to give expected frequencies greater than 5; one is subtracted because the totals of the expected and observed frequencies are constrained to be the same; and m is the number of parameters of the fitted distribution calculated from the observed data.

In example 4.4, $m = 1$, since the parameter p had to be estimated from the data.

In example 4.9, $m = 1$, since the parameter λ had to be estimated from the data.

In example 4.15, $m = 2$, since both the mean, μ, and standard deviation, σ, of the normal distribution had to be estimated from the data.

Ex. 5.16. Test the expected and observed frequencies in example 4.15 for goodness of fit.

Class mid-value x_i	e_i		o_i		$(e_i - o_i)^2/e_i$
0.3245	3.1 }	10.8	3 }	10	0.0593
0.3265	7.7		7		
0.3285	16.3		18		0.1773
0.3305	24.1		23		0.0502
0.3325	23.3		24		0.0210
0.3345	15.6		15		0.0231
0.3365	7.2 }	9.9	7 }	10	0.0010
0.3385	2.7		3		
					0.3319

From (5.26), $\nu = 6 - 1 - 2 = 3$.

Hence $\chi^2 = 0.3319$ with 3 d.f.s.

From Table A3, $\chi_3^2(0.05) = 7.81 > 0.3319$.

Hence the fit of the normal distribution to the data is satisfactory. Note that in the calculation of χ^2 the individual contributions $(e_i - o_i)^2/e_i$ are written down in the final column. This has two advantages; (a) it makes checking of the final value of χ^2 simpler, (b) if the fit is poor the classes with large contributions to the value of χ^2 will be those where the fit is worst. These may be picked out readily by glancing down the final column.

5.10 INTRODUCTION TO BAYESIAN INFERENCE

Bayesian inference is based on Bayes' Theorem (equation (3.17)). In this application we have a number of hypotheses $H_1, H_2, \ldots, H_k$ and we carry out an experiment to distinguish between these hypotheses. Let A denote the outcome of this experiment. Then Bayes' Theorem gives

$$\Pr(H_i|A) = \frac{\Pr(A|H_i)\,\Pr(H_i)}{\Pr(A|H_1)\,\Pr(H_1) + \ldots + \Pr(A|H_k)\,\Pr(H_k)} \qquad (5.27)$$

$$i = 1, 2, \ldots, k \ .$$

Here $\Pr(H_i|A)$ is referred to as the *posterior probability* of the hypothesis H_i (given the result of the experiment) and $\Pr(H_i)$ is the *prior probability* of the hypothesis H_i. We first illustrate the application of (5.27) by considering the situation of example 3.6 in a slightly different way.

Ex. 5.17. A person from a particular population may have a disease (H_1) or may not have a disease (H_2). The probability of a person chosen at random from the population having the disease is 0.02. The probability that a diagnostic test gives a positive result when the disease is present is 0.75 and when the disease is not present is 0.03. If the test gives a positive result for a person what is the probability that the person has the disease?

In this example the prior probabilities are $\Pr(H_1) = 0.02$ and $\Pr(H_2) = 1 - \Pr(H_1) = 0.98$.

Here the experiment is the diagnostic test and the result is positive (P). Then

$$\Pr(P|H_1) = 0.75$$

and

$$\Pr(P|H_2) = 0.03.$$

We require the posterior probability of H_1, $\Pr(H_1|P)$, which from (5.27) is given by

$$\Pr(H_1|P) = \frac{\Pr(P|H_1)\,\Pr(H_1)}{\Pr(P|H_1)\,\Pr(H_1) + \Pr(P|H_2)\,\Pr(H_2)}$$

$$= \frac{0.75 \times 0.02}{0.75 \times 0.02 + 0.03 \times 0.98}$$

$$= 0.338 \ .$$

This is the same result as obtained in example 3.6, but has been viewed from the point of view of evaluating posterior probabilities.

Now consider the situation in which inferences are to be made about the mean μ of a normal distribution with known variance σ^2 on the basis of a sample of n observations $x_1, x_2, \ldots, x_n$ from the distribution.

We shall assume that the prior information about μ can be expressed in the form

$$\mu \sim N(\mu_0 ; \sigma_0^2) \ . \tag{5.28}$$

That is before the experiment (making n observations) has taken place our information about μ can be condensed into the statement that it is normally distributed with a mean of μ_0 and a variance of σ_0^2.

If $\sigma_0^2 = 0$, then we know that $\mu = \mu_0$ before the experiment has taken place.

As our prior information about μ becomes less precise, the value of σ_0^2 increases. The particular case 'σ_0^2 = infinity' corresponds to the situation where we have no prior information about μ.

Let x denote the set of observations $x_1, x_2, \ldots, x_n$. Then the application of Bayes' Theorem shows (after quite a lot of algebra!) that the posterior distribution of μ given x, $(\mu|x)$, is also normal, with mean μ_n and variance σ_n^2, say; that is

$$(\mu|x) \sim N(\mu_n \; ; \sigma_n^2) \tag{5.29}$$

where

$$\mu_n = \frac{\dfrac{n\bar{x}}{\sigma^2} + \dfrac{\mu_0}{\sigma_0^2}}{\dfrac{n}{\sigma^2} + \dfrac{1}{\sigma_0^2}} \tag{5.30}$$

and

$$\sigma_n^2 = \frac{1}{\dfrac{n}{\sigma^2} + \dfrac{1}{\sigma_0^2}} \; . \tag{5.31}$$

If $\sigma_0^2 = 0$, corresponding to the prior information $\mu = \mu_0$, (5.30) and (5.31) give $\mu_n = \mu_0$ and $\sigma_n^2 = 0$; in this case, as we might have expected, the information in the sample has not changed our knowledge about μ.

On the other hand if σ_0^2 is infinite, corresponding to prior ignorance, (5.30), (5.31) and (5.29) tell us that

$$(\mu|x) \sim N\left(\bar{x} \; ; \; \frac{\sigma^2}{n}\right) \; . \tag{5.32}$$

The methods by which the result (5.32) may be used to make inferences on μ will now be considered. The posterior distribution of μ is illustrated in Fig. 5.9. The *Bayesian point estimator* $\hat{\mu}_B$ of μ is that value of $(\mu|x)$ for which the posterior probability has a maximum value. Thus

$$\hat{\mu}_B = \bar{x} \; . \tag{5.33}$$

The $100(1 - \alpha)\%$ *credible region* for μ has bounds μ_L and μ_U such that

$$\Pr(\mu_L < \mu < \mu_U | x) = 1 - \alpha$$

and also such that all the values of μ within the interval μ_L to μ_U have higher posterior probabilities than any value of μ outside the interval.

If $\alpha = 0.05$ it will be seen from Fig. 5.9 that

$$\mu_L = \bar{x} - 1.96 \frac{\sigma}{\sqrt{n}} \quad \text{and} \quad \mu_U = \bar{x} + 1.96 \frac{\sigma}{\sqrt{n}} \; . \tag{5.34}$$

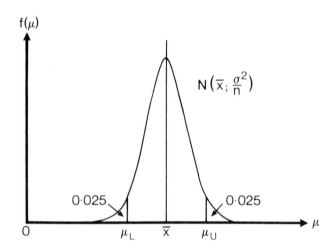

Fig. 5.9 – Posterior distribution and 95% credible region for μ.

The credible region is the Bayesian equivalent of the confidence interval introduced in section 5.2.2.

The Bayesian equivalent of hypothesis testing in this situation is considered in part (c) of the example below.

Ex. 5.18. The mean weight of a random sample of twenty tablets taken from a batch is 0.33086 g. If the weights of the tablets are known to be normally distributed with a standard deviation of 3.14×10^{-3} g

(a) find the Bayesian estimator for μ,
(b) find the 99% credible region for μ,
(c) test the hypothesis that the mean weight of the tablets in the batch is less than 0.33142 g.

Let $\mathbf{x} \, (\equiv x_1, x_2, \ldots, x_{20})$ denote the information in the sample. As no prior information is given the result (5.32) is used. Then

$$(\mu|\mathbf{x}) \sim N\left(0.33086; \frac{\left(3.14 \times 10^{-3}\right)^2}{20}\right) .$$

(a) From (5.33) $\hat{\mu}_B = 0.33086$.
(b) From (5.34) and Table A1(b) the 99% credible region for μ is

$$0.33086 \pm 2.58 \times \frac{3.14 \times 10^{-3}}{\sqrt{20}}$$

$$= 0.33086 \pm 0.00181$$

$$= 0.3290, 0.3327 .$$

(c) Let H denote the hypothesis $\mu < 0.33142$. Then

$$\Pr(H|x) = \Pr\left((\mu|x) < 0.33142)\right)$$

$$= \Pr\left(Z < \frac{0.33142 - 0.33086}{\dfrac{3.14 \times 10^{-3}}{\sqrt{20}}}\right) \text{ where } Z \sim N(0; 1)$$

$$= \Pr(Z < 0.80)$$

$$= 0.7881 .$$

We might accept H only if $\Pr(H|x)$ is greater than 0.95, 0.99 or 0.999 (corresponding to the significance levels of 0.05, 0.01 and 0.001 introduced in section 5.2.3). In this example there is not sufficient evidence to accept H.

From part (c) of the above example it will be observed that in the context of Bayesian inference a hypothesis may be considered by itself rather than as one of a pair of null and alternative hypotheses. However, there is no straightforward Bayesian equivalent of the two-sided significance test introduced in section 5.2.3.

It is possible to derive Bayesian inference methods for the situations considered in sections 5.2 to 5.8. In many of these situations the results derived using Bayesian methods are closely related to those derived using classical inference techniques when no prior information is available on the parameters concerned. However, as will be seen from equations (5.29) to (5.31) Bayesian inference methods allow prior information on the parameters to be used in conjuction with information in the sample in a natural way — a facility which is not available with the classical inference techniques considered earlier in this chapter.

5.11 BASIC PROGRAMS

In this section four programs will be presented.

(a) Program 5.1. Selection of a random sample.
     This program allows the selection of a random sample of size n, with or without replacement, from a population of size N.
(b) Program 5.2. Illustration of the sampling distribution of the mean.
     This program illustrates the sampling distribution of the mean for samples of size n from the population of Table 5.1.
(c) Program 5.3. Illustration of statistical inference.
     This program will carry out the calculations for the situation of section 5.8.1.
(d) Program 5.4. Testing the goodness of fit of a distribution.

Program listings and sample outputs
(a) Program 5.1

```
100 PRINT'□':REM**CLEAR SCREEN
110 PRINT"          **PROGRAM 5.1**"
120 PRINT"SELECTION OF A RANDOM SAMPLE OF SIZE N"
130 PRINT"   WITH, OR WITHOUT, REPLACEMENT"
140 PRINT"   FROM A POPULATION OF SIZE NO"
150 PRINT
160 DIM X(100)
170 PRINT"NO=";:INPUT NO
180 PRINT"N=";:INPUT N
190 PRINT"IS SAMPLING WITH REPLACEMENT (Y/N)";:INPUT A$
200 XO=RND(-2):REM**INITIALISES RANDOM NUMBER SEQUENCE
210 K=1
220 PRINT"VALUES FOR SAMPLE NUMBER ";K;"ARE:-"
230 FOR I=1 TO N
240 REM**GENERATE AN INTEGER IN THE RANGE 1 TO NO AT RANDOM
250 X(I)=INT(NO*RND(1))+1
260 IF A$="Y" OR I=1 THEN 340
270 REM**TEST IF INTEGER HAS BEEN SELECTED PREVIOUSLY
280 REM**IF SAMPLING IS WITHOUT REPLACEMENT
290 J=1
300 IF X(I)=X(J) THEN 250
310 J=J+1
320 IF J=I THEN 340
330 GOTO 300
340 PRINT X(I);
350 IF I=INT(I/10)*10 THEN PRINT
360 NEXT I
370 K=K+1
380 PRINT
390 PRINT"DO YOU REQUIRE A FURTHER SAMPLE (Y/N)";:INPUT B$
400 IF B$="Y" THEN 220
410 END
```

Sample of the output from program 5.1

```
POPULATION SIZE= 20
SAMPLE SIZE   = 10

IS SAMPLING WITH REPLACEMENT (Y/N)? Y

VALUES FOR SAMPLE NUMBER 1 ARE:-
 6  3  11  8  3  7  15  12  7  8

DO YOU REQUIRE A FURTHER SAMPLE (Y/N)? Y

VALUES FOR SAMPLE NUMBER 2 ARE:-
13  ι  5  7  18  2  11  2  1  8


POPULATION SIZE= 20
SAMPLE SIZE   = 10

IS SAMPLING WITH REPLACEMENT (Y/N)? N

VALUES FOR SAMPLE NUMBER 1 ARE:-
 6  3  11  8  7  15  12  13  1  5

DO YOU REQUIRE A FURTHER SAMPLE (Y/N)? Y

VALUES FOR SAMPLE NUMBER 2 ARE:-
 7  18  2  11  1  8  17  5  10  12
```

(b) Program 5.2

```
100 PRINT"◻":REM**CLEAR SCREEN
110 PRINT"      **PROGRAM 5.2**"
120 PRINT"ILLUSTRATION OF THE SAMPLING"
130 PRINT"  DISTRIBUTION OF THE MEAN"
140 PRINT"  FOR 1000 SAMPLES OF SIZE N"
150 PRINT" FROM THE DATA OF TABLE 5.1"
160 PRINT
170 DIM X(100),F(100)
180 FOR I=1 TO 100:READ X(I):NEXT I
190 X0=RND(-3):REM**INITIALISE RANDOM NUMBER SEQUENCE
200 REM**CALCULATION OF POP. MEAN,MP, AND VAR. VP
210 S1X=0 : S2X2=0
220 FOR I=1 TO 100
230 S1X=S1X+X(I)
240 S2X2=S2X2+X(I)*X(I)
250 NEXT I
260 MP=S1X/100
270 VP=(S2X2-S1X↑2/100)/100
280 PRINT"SAMPLE SIZE(N)=";:INPUT N
290 VS=VP/N
300 PRINT
310 PRINT"THEORETICAL VALUES OF THE MEAN"
320 PRINT"AND VARIANCE OF THE SAMPLING"
330 PRINT"DISTRIBUTION OF THE MEAN ARE"
340 PRINT MP;" AND ";VS;" WHEN N IS ";N
350 REM**CALCULATE REASONABLE CLASS WIDTH
360 REM**R FOR A SAMPLE OF SIZE N
370 R=INT(200/SQR(N)+0.5)/10
380 REM**CALCULATE LOWER BOUND OF FIRST CLASS
390 LB=INT(10*(MP-6*SQR(VP)))/10
400 IF LB<0 THEN LB=0
410 REM**ZERO SUMS AND FREQUENCIES
420 S3M=0:S4M2=0:FOR I=1 TO 100:F(I)=0:NEXT I
430 REM**MEANS OF 1000 SAMPLES OF SIZE N
440 FOR J=1 TO 1000
450 S5=0
460 FOR I=1 TO N
470 REM**CHOOSE RANDOM K BETWEEN 1 AND 100
480 K=100*RND(1)+1
490 S5=S5+X(K)
500 NEXT I
510 REM**CALCULATE SAMPLE MEAN
520 SM=S5/N
530 S3M=S3M+SM
540 S4M2=S4M2+SM*SM
550 REM**REDUCE MEANS TO A FREQUENCY DISTRIBUTION
560 K1=INT((SM-LB)/R)+1
570 F(K1)=F(K1)+1
580 NEXT J
590 REM**ESTIMATES OF THE MEAN AND VARIANCE
600 REM**OF THE SAMPLING DISTRIBUTION OF THE MEAN
610 M3=S3M/1000
620 V3=(S4M2-S3M↑2/1000)/999
630 PRINT
640 PRINT"SIMULATION ESTIMATES OF THE"
650 PRINT"MEAN AND VARIANCE OF THE"
660 PRINT"SAMPLING DISTRIBUTION OF THE"
670 PRINT"MEAN ARE ";M3;" AND ";V3
680 PRINT
690 PRINT"PRESS ANY KEY TO CONTINUE"
700 GET A$ : IF A$="" THEN 700
710 PRINT"◻":REM**CLEAR SCREEN
720 PRINT
```

```
730 REM**CALCULATE MID-VALUE OF FIRST CLASS
740 X2=LB+R/2
750 PRINT"CLASS MID-VALUE   FREQUENCY"
760 FOR I=1 TO 50
770 IF F(I)=0 THEN 790
780 PRINT TAB(5);X2;TAB(21);F(I)
790 X2=X2+R
800 NEXT I
810 PRINT
820 PRINT"DO YOU WANT TO REPEAT THE EXPERIMENT(Y/N)";
830 INPUT B$ : IF B$="Y" THEN 280
840 DATA 67,53,70,47,46,61,63,23,125,10
850 DATA 37,7,21,11,25,72,112,145,120,2
860 DATA 87,35,18,18,32,26,54,114,59,112
870 DATA 64,17,4,197,36,3,225,15,62,37
880 DATA 13,31,63,10,46,63,40,113,4,48
890 DATA 30,69,81,0,14,81,96,44,29,32
900 DATA 58,4,3,63,26,30,21,131,67,40
910 DATA 224,51,65,68,20,145,16,10,12,50
920 DATA 23,101,40,9,0,27,63,18,0,56
930 DATA 51,39,31,118,56,17,166,141,11,82
940 END
```

Sample of the output from program 5.2

```
SAMPLE SIZE(N)= 4

THEORETICAL VALUES OF THE MEAN
AND VARIANCE OF THE SAMPLING
DISTRIBUTION OF THE MEAN ARE
 53.92  AND   562.693399  WHEN N IS   4

SIMULATION ESTIMATES OF THE
MEAN AND VARIANCE OF THE
SAMPLING DISTRIBUTION OF THE
MEAN ARE  53.97725  AND  611.954372


CLASS MID-VALUE  FREQUENCY
      5              3
     15             33
     25            117
     35            175
     45            185
     55            143
     65            104
     75             87
     85             56
     95             45
    105             27
    115             18
    125              2
    135              1
    145              1
    155              1
    165              2
```

(c) Program 5.3

```
100 PRINT"⊔":REM**CLEAR SCREEN
110 PRINT"           **PROGRAM 5.3**"
120 PRINT"   STATISTICAL INFERENCES ON THE"
130 PRINT"DIFFERENCE OF MEANS MU1 AND MU2 OF"
140 PRINT"      TWO NORMAL POPULATIONS"
150 PRINT
160 REM**INPUT DATA,CALCULATE SAMPLE MEANS M1 AND M2,  AND
170 REM**ESTIMATE VHAT OF COMMON VARIANCE
180 M1=0 : M2=0 : C1=0 : C2=0
190 PRINT"INPUT SAMPLE SIZES N1 AND N2"
200 PRINT"N1=";:INPUT N1
210 PRINT"N2=";:INPUT N2
220 PRINT"INPUT OBSERVATIONS FROM THE FIRST SAMPLE"
230 FOR I=1 TO N1
240 PRINT"X1(";I;")=";:INPUT X1
250 M1=M1+X1
260 C1=C1+X1*X1
270 NEXT I
280 PRINT"INPUT OBSERVATIONS FROM THE SECOND SAMPLE"
290 FOR I=1 TO N2
300 PRINT"X2(";I;")=";:INPUT X2
310 M2=M2+X2
320 C2=C2+X2*X2
330 NEXT I
340 REM**CALCULATE SUMS OF SQUARES CORRECTED FOR THE MEAN
350 C1=C1-M1*M1/N1
360 C2=C2-M2*M2/N2
370 REM**CALCULATE MEANS
380 M1=M1/N1
390 M2=M2/N2
400 PRINT
410 PRINT"CONFIDENCE COEFFICIENT=";:INPUT TC
420 PRINT"CONFIDENCE LEVEL(PERCENT)=";:INPUT A
430 VHAT=(C1+C2)/(N1+N2-2)
440 D=SQR(VHAT*(1/N1+1/N2))
450 REM**ESTIMATE OF DIFFERENCE OF MEANS
460 MHAT=M1-M2
470 REM**CONFIDENCE LIMITS
480 LC=MHAT-TC*D
490 LC=INT(1000*LC+0.5)/1000
500 UC=MHAT+TC*D
510 UC=INT(1000*UC+0.5)/1000
520 REM**STUDENT'S T
530 T=MHAT/D
540 PRINT"⊔":REM**CLEAR SCREEN
550 PRINT"ESTIMATE OF MU1-MU2 IS ";MHAT
560 PRINT
570 PRINT A;"PERCENT CONFIDENCE LIMITS ARE ";LC;" AND ";UC
580 PRINT
590 PRINT"TEST STATISTIC FOR THE NULL"
600 PRINT"HYPOTHESIS MU1=MU2 IS"
610 PRINT"T=";INT(1000*T+0.5)/1000;" ON ";N1+N2-2;" D.F.'S"
620 PRINT:PRINT
630 END
```

Sample of the output from program 5.3

```
INPUT SAMPLE SIZES N1 AND N2
N1= 4
N2= 5
INPUT OBSERVATIONS FROM THE FIRST SAMPLE
X1( 1 )= 1.8
X1( 2 )= .9
X1( 3 )= 1.3
X1( 4 )= 1.5
INPUT OBSERVATIONS FROM THE SECOND SAMPLE
X2( 1 )= 2.4
X2( 2 )= 2.8
X2( 3 )= 2
X2( 4 )= 1.9
X2( 5 )= 2.2

CONFIDENCE COEFFICIENT= 2.365
CONFIDENCE LEVEL(PERCENT)= 95

ESTIMATE OF MU1-MU2 IS -.885

 95 PERCENT CONFIDENCE LIMITS ARE -1.466  AND -.304

TEST STATISTIC FOR THE NULL
HYPOTHESIS MU1=MU2 IS
T=-3.601  ON  7  D.F.'S
```

(d) Program 5.4

```
100 PRINT"□":REM**CLEAR SCREEN
110 PRINT"          **PROGRAM 5.4**"
120 PRINT"        COMPARISON OF A SET OF"
130 PRINT"   OBSERVED (O) AND EXPECTED (E)"
140 PRINT"FREQUENCIES USING A CHI-SQUARE TEST"
150 PRINT
160 DIM O(50),E(50),O1(50),E1(50)
170 PRINT"NUMBER OF CLASSES K";:INPUT K
180 PRINT
190 PRINT"INPUT OBSERVED FREQUENCIES:-"
200 FOR I=1 TO K
210 PRINT"O(";I;")=";:INPUT O(I)
220 NEXT I
230 PRINT"INPUT EXPECTED FREQUENCIES:-"
240 FOR I=1 TO K
250 PRINT"E(";I;")=";:INPUT E(I)
260 NEXT I
270 PRINT"□":REM**CLEAR SCREEN
280 PRINT"NUMBER OF PARAMETERS M ESTIMATED FROM THE DATA=";:INPUT M
290 REM**COMBINATION OF EXPECTED FREQUENCIES
300 REM**OF LESS THAN 5 TO GIVE NEW ARRAYS
310 REM**O1(I) AND E1(I)
320 JL=1
330 E1(1)=E(1):O1(1)=O(1)
340 IF E1(JL)>5 AND E(JL+1)>5 THEN 380
350 E1(JL+1)=E1(JL)+E(JL+1)
360 O1(JL+1)=O1(JL)+O(JL+1)
370 JL=JL+1:GOTO 340
380 JU=K
```

```
390 E1(K)=E(K):O1(K)=O(K)
400 IF E1(JU)>5 AND E(JU-1)>5 THEN 440
410 E1(JU-1)=E1(JU)+E(JU-1)
420 O1(JU-1)=O1(JU)+O(JU-1)
430 JU=JU-1:GOTO 400
440 FOR I=JL+1 TO JU-1
450 E1(I)=E(I)
460 O1(I)=O(I)
470 NEXT I
480 PRINT
490 REM**CALCULATION OF CHI-SQUARE
500 PRINT"CLASS   O(I)    E(I) (O(I)-E(I))↑2/E(I)"
510 CHISQ=0
520 FOR I=JL TO JU
530 D=(O1(I)-E1(I))↑2/E1(I)
535 D1=INT(1000*D+0.5)/1000
540 PRINT TAB(1);I;TAB(7);O1(I);TAB(13);E1(I);TAB(23);D1
550 CHISQ=CHISQ+D
560 NEXT I
570 NU=JU-JL-M
580 PRINT
590 PRINT"CHISQ=";INT(1000*CHISQ+0.5)/1000;" ON";NU;" D.F.'S"
600 PRINT
610 END
```

Output of program 5.4 for the data of example 5.15

```
NUMBER OF CLASSES K= 8

INPUT OBSERVED FREQUENCIES:-
O( 1 )= 3
O( 2 )= 7
O( 3 )= 18
O( 4 )= 23
O( 5 )= 24
O( 6 )= 15
O( 7 )= 7
O( 8 )= 3
INPUT EXPECTED FREQUENCIES:-
E( 1 )= 3.1
E( 2 )= 7.7
E( 3 )= 16.3
E( 4 )= 24.1
E( 5 )= 23.3
E( 6 )= 15.6
E( 7 )= 7.2
E( 8 )= 2.7
NUMBER OF PARAMETERS M ESTIMATED FROM THE DATA= 2
```

CLASS	O(I)	E(I)	$(O(I)-E(I))↑2/E(I)$
2	10	10.8	.059
3	18	16.3	.177
4	23	24.1	.05
5	24	23.3	.021
6	15	15.6	.023
7	10	9.9	1E-03

```
CHISQ= .332  ON 3  D.F.'S
```

APPENDIX 5.1 DERIVATION OF THE $100(1 - \alpha)\%$ CONFIDENCE LIMITS FOR μ

We have, with probability $(1 - \alpha)$

$$-z(\alpha/2) < Z < z(\alpha/2)$$

hence, with probability $(1 - \alpha)$

$$-z(\alpha/2) < \frac{\bar{X} - \mu}{\sigma/\sqrt{n}} < z(\alpha/2) \ .$$

Considering these inequalities separately we have firstly

$$-z(\alpha/2) < \frac{\bar{X} - \mu}{\sigma/\sqrt{n}} \ .$$

Multiplying by $\sigma/\sqrt{n}$ gives

$$-z(\alpha/2) \frac{\sigma}{\sqrt{n}} < \bar{X} - \mu$$

adding $\mu + z(\alpha/2) \dfrac{\sigma}{\sqrt{n}}$ to both sides of the inequality leads to

$$\mu < \bar{X} + z(\alpha/2) \frac{\sigma}{\sqrt{n}} \ .$$

Similarly the second inequality leads to

$$\bar{X} - z(\alpha/2) \frac{\sigma}{\sqrt{n}} < \mu,$$

and combining these results gives

$$\bar{X} - z(\alpha/2) \frac{\sigma}{\sqrt{n}} < \mu < \bar{X} + z(\alpha/2) \frac{\sigma}{\sqrt{n}} \ .$$

Replacing $\bar{X}$ by its sample value $\bar{x}$ leads to (5.3) and (5.4).

APPENDIX 5.2 DERIVATION OF THE $100(1 - \alpha)\%$ CONFIDENCE LIMITS FOR σ^2

Since $U = (n - 1) \dfrac{s^2}{\sigma^2} \sim \chi^2_{n-1}$ we have, with probability $(1 - \alpha)$

$$\chi^2_{n-1}(1 - \alpha/2) < (n - 1) \frac{s^2}{\sigma^2} < \chi^2_{n-1}(\alpha/2) .$$

Inverting the inequalities, which changes their sign, leads to

$$\frac{1}{\chi^2_{n-1}(1 - \alpha/2)} > \frac{\sigma^2}{(n - 1)s^2} > \frac{1}{\chi^2_{n-1}(\alpha/2)} .$$

Multiplying by $(n - 1)s^2$, and putting the lower limit at the left-hand side, gives

$$\frac{(n - 1)s^2}{\chi^2_{n-1}(\alpha/2)} < \sigma^2 < \frac{(n - 1)s^2}{\chi^2_{n-1}(1 - \alpha/2)}$$

which is the result used in example 5.8, where the sample value is substituted for the statistic s^2.

EXERCISES

Section A

1. The blood alcohol concentration required to cause respiratory failure in rats is distributed with a mean of 9.30 mg/ml and a standard deviation of 0.15 mg/ml. If samples of 36 rats are taken in an experiment and the blood alcohol concentration required to cause respiratory failure is measured, find the sampling distribution of the sample mean blood alcohol concentration.

   In what proportion of the samples would you expect the mean (a) to lie between 9.30 and 9.35 mg/ml; (b) to be greater than 9.34 mg/ml; (c) to be less than 9.31 mg/ml; (d) to lie between 9.20 and 9.40 mg/ml?

2. A random sample of 40 aspirin tablets is found to have a mean of 0.3313 g with a standard deviation of 3.20×10^{-3} g. Find (a) 95%, (b) 99.9% confidence limits for the mean weight of the tablets in the population from which the sample is drawn.

3. When a process is under control the mean weight of capsules filled by a machine is 350.0 mg. A random sample of 40 capsules is chosen from the output of the machine at a particular time and is found to have a mean weight of 353.1 mg with a standard deviation of 7.1 mg. Is the mean of the process under control at that time?

4. The mean specific gravity of sulphuric acid made by a certain process is 1.840 with a standard deviation of 0.030. A change is made in the process which is thought might lead to an increase in the mean. 100 samples were taken at random and the mean density of the 100 samples was found to be 1.844. Test whether the process mean has been increased by the change, assuming that the s.d. has not.

5. A laboratory technician finds the mean weight of a random sample of 40 aspirin tablets to be 0.3313 g with a standard deviation of 3.20×10^{-3} g. He then finds in his notebook the weights of a further five tablets which he has recorded but not taken into account in his calculations. If these weights are

   0.330   0.333   0.329   0.337   0.334

   find the best estimate you can for the population mean weight of the tablets and a 95% confidence interval for the mean.

6. Glucose treatment is expected to lengthen the sleep time of mice. The mean hexobarbital sleep time of a sample of 60 untreated mice is 28.5 min with a standard deviation of 9.0 min. The mean hexobarbital sleep time of 30 treated mice is found to be 49.2 min with a standard deviation of 10.0 min. Does this result support the hypothesis?
   Find 95% confidence limits for the increase in sleep time.

7. Of a sample of 100 operatives in a particular factory, 15 are found to have lung conditions rated as poor. Is the proportion significantly greater than 0.1? Find 95% confidence limits for the proportion of operatives in the factory with poor lung conditions.

8. The weights, in mg, of a random sample of ten aspirin tablets from a large batch are

   336, 333, 335, 333, 329, 334, 324, 331, 332, 332.

   Find 95% confidence limits for the variance of the tablets in the batch. Test whether this variance is consistent with a value of 10 mg^2.
   The weights, in mg, of a random sample of twelve aspirin tablets from a second batch are

   332, 336, 337, 336, 333, 335, 332, 334, 332, 334, 340, 332.

   Test whether the variances of the weights in the two batches are different. Test whether the mean weight of the tablets in each of batches may be taken to be 331.4 mg. Assuming that the variances of the weights in the two batches are the same, test whether the mean weights of the tablets in the two batches differ. State any assumptions made about the data in carrying out the tests.

9. It is thought that the growth of firebent is inhibited by bracken litter. The shoot lengths of firebent grown under controlled conditions is known to have a mean of 16.50 mm. The results for ten specimens grown under the same conditions with the addition of bracken litter are (in mm)

    17.5 13.0 14.0 10.0 16.5 13.5 12.0 17.0 14.0 12.5

    Do these results support the hypothesis?
    Find 95% confidence limits for the mean length of the bracken shoots grown in the litter.

10. Glucose treatment is expected to lengthen the sleep time of mice. The mean hexobarbital sleep time of 20 treated mice is found to be 49.2 min with a standard deviation of 8.2 min. Does this result support the hypothesis
    (a) if it is known that the mean sleep time of untreated mice is 28.5 min?
    (b) if it is known that the mean sleep time of untreated mice is 28.5 min with a standard deviation of 7.5 min?
    (c) if it is known only that the mean sleep time of a sample of 15 untreated mice is 28.5 min with a standard deviation of 7.25 min?

    (In (b) you may assume that the standard deviation is unaffected by the glucose treatment.)

11. Eight rabbits were given equal doses of insulin and two methods A and B were used to determine the amount of glycogen (in mls/l) in their muscles. The results were as follows

Rabbit	1	2	3	4	5	6	7	8
A	1.8	2.0	2.2	1.9	2.3	1.6	2.1	2.1
B	2.0	1.6	1.1	1.4	1.9	1.6	2.3	2.0

    Is there a significant difference between the methods?

12. Test the goodness of fit of the distributions fitted in Exercises 4A, examples 4, 7 and 15.

13. The weights of aspirin tablets in a large batch are known to be normally distributed with a standard deviation of 3.2 mg and unknown mean μ mg. A random sample of 20 tablets from the batch have a mean weight of 335.2 mg. Write down the posterior distribution for μ, assuming no prior knowledge. Find a 95% credible region for the mean weight of the tablets in the batch. Test the hypothesis that this mean weight is greater than 330 mg.

Section B
1. The strength of paper used by a company is a random variable with a mean of 2.100 kg per sq mm and a standard deviation of 0.200 kg per sq mm.

Write down the sampling distribution of the mean strength of samples of 36 pieces of paper. What proportion of sample means lie between (a) 2.030 and 2.170 kg per sq mm, (b) are greater than 2.184 kg per sq mm, (c) are less than 2.000 kg per sq. mm?

2. Thirty measurements of the percentage of iron in a compound have a mean of 12.0% with a standard deviation of 0.5%. Find 95% confidence limits for the percentage of iron in the compound.

3. A sample of 36 pieces of thread show a mean breaking strength of 190 g with a standard deviation of 30 g. Test the hypothesis that the breaking strength of the thread is less than 200 g.

4. A laboratory assistant finds the mean weight of a sample of 40 pottery vessels to be 0.3313 kg with a standard deviation of 0.0032 kg. He then finds in his notebook the weights of a further five vessels which he has recorded but not taken into account in his calculation. If these weights are (in kg)

   0.330   0.333   0.329   0.337   0.334

   find the best estimate from the data of the population mean weight for the pottery vessels and a 95% confidence interval for the mean.

5. The caesium contents of 40 samples of obsidian from source A have a mean of 3.85 ppm with a standard deviation of 0.30 ppm while the corresponding figures for 35 samples from a source B are 2.27 ppm and 0.54 ppm respectively. Is the caesium content of the obsidian from the two sources significantly different? A further 30 samples of obsidian have a mean caesium content of 4.28 ppm with a standard deviation of 0.41 ppm. Are these samples likely to have come from either of the sources A or B?

6. Before improvement to a certain section of road it was found that 40% of the traffic was travelling at less than 40 m.p.h. Out of a sample of 50 vehicles observed after the improvement, 15 are observed to have speeds of less than 40 m.p.h. Does this figure show a significant decrease?

7. Out of 300 forms sent out in a postal survey 85 are returned. Past experience has shown that a return of 25% should be expected. Is the proportion obtained in the present survey significantly different?

8. The standard deviation of the measurements made by an instrument is normally 0.11 mm. A series of 30 measurements made by the instrument exhibit a standard deviation of 0.127 mm. Is the increase significant?

9. Two classes independently measured the static coefficient of friction of brass on wood μ_{BW} and their results are summarised in the following table

μ_{BW}	0.41	0.43	0.45	0.47	0.49	0.51
f (Class I)	2	10	18	16	9	5
f (Class II)	4	16	20	17	3	4

Are the sample variances significantly different?

10. A machine produces cigarettes whose weights, when the production process is under control, are normally distributed about a mean of 1.200 g with standard deviation of 0.063 g. What proportion of cigarettes would you expect to have weights of (a) less than 1.050 g, (b) between 1.100 and 1.300 g?

   A random sample of 25 cigarettes is taken from the output to test whether the process is under control and is found to have a mean weight of 1.180 g with a standard deviation of 0.070 g. Do these results indicate that there has been (a) a change in the variance of the weights of the cigarettes produced by the machine, (b) a change in the mean weight of the cigarettes produced by the machine?

11. Twenty measurements of the palladium content of specimens of late bronze age pottery give the following results (in ppm)

   10.50 10.00 9.71 9.19   8.39 9.30 8.27 7.37 6.56 9.47
   7.19   8.67 7.60 9.01 10.72 7.80 8.02 6.92 8.74 7.58

   Test whether these measurements could have come from a population with mean 8.40 ppm.

   A second sample of twenty measurements has a mean of 8.40 ppm with a standard deviation of 1.20 ppm. Could the two samples have been drawn from populations with the same mean?

12. In a laboratory, six concrete specimens were made to a certain mix specification and their tensile strengths measured, leading to the following results

   17.6   18.1   17.5   17.7   18.0   18.0

   A further six specimens were made to the same mix specifications except that a rapid hardening additive was used. The tensile strengths of these specimens were

   17.8   18.3   18.6   17.8   17.5   18.5

   Find 95% confidence limits for the difference in strengths of the concrete with and without the additive.

   Do the results indicate that the additive has led to a change in the tensile strength?

   State any assumptions which have to be made about the data in order to carry out the necessary statistical tests.

13. Five students determine the percentage of a component in a mixed fluid each using two methods A and B. The results obtained by the students are tabulated below

Student	1	2	3	4	5
Method A	21.4	22.7	22.0	21.6	21.8
Method B	21.8	22.2	23.2	22.1	22.2

Test whether these results indicate a difference between the methods A and B.

14. Test the goodness of fit of the distributions fitted in Exercises 4B, examples 6, 11 and 20.

15. The lifetime of a component is known to be normally distributed with a standard deviation of 50 hours and unknown mean μ hours. A random sample of 20 components has a mean lifetime of 490 hours. Write down the posterior distribution for μ, assuming no prior knowledge. Find a 95% credible region for μ and test the hypothesis that μ is less than 500.

Section C†

1. Write programs to give (a) estimates, (b) confidence limits, and (c) values of the test-statistics for the inference situations discussed in sections 5.2, 5.3, 5.4, 5.5, 5.6, 5.7 and 5.8.2.

2. Using program 4.3 to generate samples from appropriate normal distributions, apply the programs developed in exercise 1 above.

3. Adapt program 4.1 to find the sampling distribution of the mean of samples of size n (> 30) from a binomial distribution. Use program 4.3 to fit a normal distribution to this data and program 5.4 to test the goodness of fit of the observed and expected frequencies.

4. Combine programs 4.2 and 5.4 to generate a sample from a Poisson distribution and test the goodness of fit of the generated and theoretical frequencies.

†Check the operation of the programs developed in this section on data from the examples of sections A and B, or worked examples in the text.

6

Regression and Correlation

6.1 INTRODUCTION

In this chapter we shall consider the problem of fitting the 'best straight line' through a set of points $(x_1, y_1), (x_2, y_2), \ldots, (x_n, y_n)$ which do not lie exactly on a line. The set of points may not lie exactly on a line because

(i) the underlying relationship between x and y is (at least for all practical purposes) exactly linear, but the observation y may be subject to a random experimental error. Consider an experiment in which the extension y mm of a spring, subject to a load x g, is measured using a metre stick. The law relating y and x is linear, but the experimental points will be scattered about a line because of the crude method used to measure y; here it would be relatively straightforward to eliminate the experimental error by measuring y using a vernier;

(ii) the underlying relationship between the variables is only approximately linear and no amount of refinement in the measurement technique will make the points (x, y) lie exactly on a straight line; for example y mm Hg might be a man's blood pressure and x years his age.

For historical reasons the straight line of best fit drawn through a set of points is called the *regression line,* and the gradient of the line is referred to as the *coefficient of regression.* This line is also called the *least squares line* through the set of points, since the line is fitted using the method of *least squares.*

In considering the line of regression of y on x, y will be referred to as the *dependent* or *response* variable, and x will be referred to as the *explanatory* variable (x may also be referred to as the *controlled,* or *independent* variable).

In practice three types of linear relationship between two variables can arise.

(a) The variable x can be controlled with negligible error by the experimenter; the random experimental error is associated with the dependent variable Y. This is the situation considered in (i) above.

(b) The variables X and Y may be jointly randomly distributed. For example X kg could be the weight of an adult male and Y m his height.
(c) Both X and Y are subject to independent random experimental errors, neither of which can be neglected. For example X m/s could be the velocity of a liquid flowing down a pipe measured by one type of meter and Y m/s could be the same velocity measured by a second type of meter.

Situation (c) is rather more complex than (a) or (b) and will not be considered here.

6.2 FITTING A REGRESSION LINE WHEN THE EXPLANATORY VARIABLE IS CONTROLLED

We consider first the situation in which the explanatory variable x is *controlled*. This means that we may set the value of x effectively without error, and that if the experiment was repeated we could use the same set of x values; thus x could be a weight, or quantity of a drug, or the time from the start of the experiment. The model adopted for the response Y_j $(j = 1, \ldots, n)$ corresponding to $x = x_j$ is

$$Y_j = \alpha + \beta x_j + \epsilon_j \qquad (j = 1, \ldots, n) \quad . \qquad (6.1)$$

Here α and β are (unknown) constants and ϵ_j represents the random experimental error. We assume that the ϵ_js are independently and identically distributed with mean zero and variance σ^2. Thus if we kept x_j fixed and measured Y_j a large number of times we would find that the Y_j values would be distributed about a mean of $\alpha + \beta x_j$ with a variance of σ^2. To apply the technique of analysis of regression (section 6.2.1) it is necessary to make the further assumption that the ϵ_js (and hence the Y_js) are normally distributed.

Statistical tests can be carried out to verify that given data follows a model with the above properties. However, the details of these tests will not be considered in this chapter.

The type of data under consideration is illustrated in Table 6.1 relating to the ability of liver slices from guinea pigs of different ages to conjugate phenolphthalein with glucoronic acid. Here x is the age of the guinea pig in days and y is the number of nanomoles of phenolphthalein conjugated.

Table 6.1

x	1	1	3	5	6	10	10	11	14	15	21
y	6	9	12	18	31	38	44	22	37	46	54

The x variable here is under the control of the experimenter and the experiment could be repeated if necessary using the same set of x values. Note however that for repeat measurements at a given value of x we do not necessarily obtain the same value of y; for example when $x = 10$ (twice) we find $y = 38$ and $y = 44$.

The first step in the analysis of the results is to draw a *scatter diagram* (or graph) of y against x (Fig. 6.1). It is evident from the scatter diagram that there is an approximate linear relationship between x and y. We could attempt to judge the best straight line through these points by eye, but using this approach different people would obtain different 'best' lines. Furthermore we should be unable to test whether the resulting linear relationship was statistically significant; that is whether the amount of variation in the values y explained by the linear relationship is large in comparison with the scatter of the y values about the line.

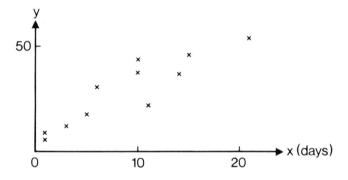

Fig. 6.1 – Scatter diagram for the data of Table 6.1.

To overcome these difficulties we use the criterion of *least squares* to find the equation of the best straight line through the points. Imagine a straight line $y = a + bx$ drawn through the points.

Consider S the sum of squares of the deviations e_i of the responses y_i at x_i from the corresponding points $\hat{y}_i = a + bx_i$ on the estimated line (Fig. 6.2). Then

$$S = \sum_{i=1}^{n} e_i^2$$

$$= \sum_{i=1}^{n} (y_i - \hat{y}_i)^2$$

$$= \sum_{i=1}^{n} (y_i - a - bx_i)^2 \tag{6.2}$$

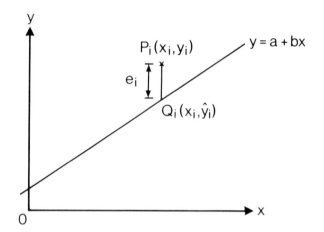

Fig. 6.2 – Deviations from the least squares line.

The *least squares estimates* a and b of α and β are found by minimising S. This is done by equating the partial derivatives $\partial S/\partial a$ and $\partial S/\partial b$ to zero. The resulting equations may be solved for a and b leading to (see Appendix 6.1)

$$b \;=\; \frac{C_{xy}}{C_{xx}} \tag{6.3}$$

and

$$a \;=\; \bar{y} - b\bar{x} \tag{6.4}$$

Here

$$C_{xx} \;=\; \sum_{i=1}^{n} (x_i - \bar{x})^2 \tag{6.5a}$$

$$=\; \sum_{i=1}^{n} x_i^2 - \frac{1}{n}\left(\sum_{i=1}^{n} x_i\right)^2 \tag{6.5b}$$

and

$$C_{xy} \;=\; \sum_{i=1}^{n} (x_i - \bar{x})(y_i - \bar{y}) \tag{6.6a}$$

$$=\; \sum_{i=1}^{n} x_i y_i - \frac{1}{n}\left(\sum_{i=1}^{n} x_i\right)\left(\sum_{i=1}^{n} y_i\right) \;. \tag{6.6b}$$

The statistic $C_{xy}/(n-1)$ is defined to be the sample *covariance* of x and y Some algebra and the definitions of $\bar{x}$ and $\bar{y}$ are needed to obtain the alternative forms of C_{xx} and C_{xy} (see Appendix 6.2). When carrying out the calculations with the aid of an electronic calculator formulae (6.5b) and (6.6b) are generally the most convenient to use. Numerical difficulties can arise, however, when

$$\sum_{i=1}^{n} x_i^2 \text{ is nearly equal to } \frac{1}{n} \left(\sum_{i=1}^{n} x_i \right)^2$$

and

$$\sum_{i=1}^{n} x_i y_i \text{ is nearly equal to } \frac{1}{n} \left(\sum_{i=1}^{n} x_i \right) \left(\sum_{i=1}^{n} y_i \right) .$$

In these circumstances a large number of significant figures can be lost in taking the differences of the two terms. However, this behaviour will become evident in the course of the calculation and then resort can be made to formulae (6.5a) and (6.6a), or coding can be used (Appendix 6.4).

If a computer is used it will not be evident when the two terms in either (6.5b) or (6.6b) are nearly equal and inaccurate values for a and b could be obtained without any indication that this is the case. Under these circumstances it is probably better to play safe by first calculating $\bar{x}$ and $\bar{y}$ and then making use of the formulae (6.5a) and (6.6a).

The application of these results will be illustrated by consideration of the data relating to the conjugation of phenolphthalein with glucoronic acid in guinea pig livers (Table 6.1).

Ex. 6.1. Use the method of least squares to fit a straight line to the data tabulated below. Illustrate the line and data on the same graph.

x	1	1	3	5	6	10	10	11	14	15	21
y	6	9	12	18	31	38	44	22	37	46	54

Estimate the value of y when $x = 8$.

For the above data

$$n = 11, \quad \Sigma x = 97, \quad \Sigma x^2 = 1255, \quad \Sigma y = 317, \quad \Sigma xy = 3731 \quad .$$

Then $\bar{x} = 8.8182$ $\qquad C_{xx} = 1255 - \dfrac{1}{11} \times 97^2 = 399.6364$

$\bar{y} = 28.8182$ $\qquad C_{xy} = 3731 - \dfrac{1}{11} \times 317 \times 97 = 935.6364 \quad .$

From (6.3) $b = \dfrac{935.6364}{399.6364} = 2.3412.$

From (6.4) $a = \bar{y} - b\bar{x} = 8.1729.$

Then the least squares line has equation

$$y = 8.17 + 2.34x .$$

When $x = 8$

$$\hat{y}_8 = 8.17 + 2.34 \times 8 = 26.89 \doteqdot 27 .$$

To reduce rounding-off errors a large number of significant figures are kept during the calculation, but at the end the coefficients a and b are rounded off to a meaningful number of significant figures.

The data and the regression line are illustrated in Fig. 6.3.

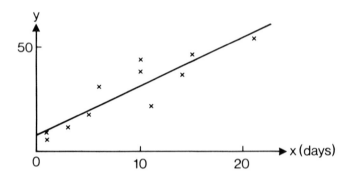

Fig. 6.3 – Least squares regression line for the data of Table 6.1.

6.2.1 Analysis of regression

The total variation of the experimentally measured responses $y_i (i = 1, \ldots , n)$ about their mean value $\bar{y}$ is

$$C_{yy} = \sum_{i=1}^{n} (y_i - \bar{y})^2 \ (= \sum_{i=1}^{n} y_i^2 - \frac{1}{n} (\Sigma y_i)^2) . \tag{6.7}$$

The deviation of y_i from $\bar{y}$ may be split into two components, $y_i - \hat{y}_i$ which represents the deviation of the experimental point from the corresponding point on the regression line, and $\hat{y}_i - \bar{y}$ which represents the component of the deviation explained by the regression of y on x (Fig. 6.4). Thus

$$y_i - \bar{y} = (y_i - \hat{y}_i) + (\hat{y}_i - \bar{y}) \tag{6.8}$$

where

$$\hat{y}_i = a + bx_i .$$

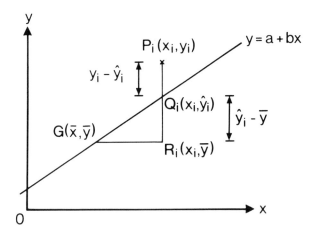

Fig. 6.4 — Analysis of regression.

Correspondingly C_{yy} may be split up into two separate sums of squares

$$C_{yy} = \sum_{i=1}^{n} (y_i - \hat{y}_i)^2 + \sum_{i=1}^{n} (\hat{y}_i - \bar{y})^2 \tag{6.9}$$

(see Appendix 6.3).

The second sum of squares is

$$\sum_{i=1}^{n} (\hat{y}_i - \bar{y})^2 = \sum_{i=1}^{n} (a + bx_i - \bar{y})^2$$

$$= \sum_{i=1}^{n} (bx_i - b\bar{x})^2, \text{ using } \bar{y} = a + b\bar{x}$$

$$= b^2 \sum_{i=1}^{n} (x_i - \bar{x})^2$$

$$= b^2 C_{xx} \quad (= C_{xy}^2/C_{xx}) \quad . \tag{6.10}$$

This is the sum of squares due to regression (SSR) and gives the part of the total variation of y about its mean value accounted for by the regression of y on x.

The *residual* $e_i = \hat{y}_i - y_i$ is the deviation of y_i from its estimated value $\hat{y}_i$, and the first sum of squares in (6.9)

$$\sum_{i=1}^{n} (y_i - \hat{y}_i)^2$$

measures the part of the total variation of y due to scatter about the regression line; it is referred to as the residual, or error, sum of squares (SSE).

The technique of splitting the total sum of squares about the mean (C_{yy}) into separate components is called *analysis of variance.*

If the experiment was repeated a large number of times using the same set of x values each time and SSE calculated for each of the experiments then it can be shown from theoretical considerations (using the properties of the random error term ϵ_j postulated at the beginning of this section) that the average value of SSE would be

$$(n - 2)\sigma^2$$

where $\sigma^2 = \text{var}(\epsilon_j)$. Hence we can obtain an estimate of σ^2 from

$$\hat{\sigma}^2 = \frac{1}{n-2} \; \text{SSE}$$

$$= \frac{1}{n-2} (C_{yy} - b^2 C_{xx}) \tag{6.11}$$

where we have used $C_{yy} = \text{SSR} + \text{SSE}$.

The analysis of variance (ANOVA) table for testing the significance of the regression is set out below.

ANOVA table

Source of variation	d.f	Sum of squares	Mean square[1]
Regression	1[2]	SSR	SSR
Residual	$n-2$[3]	SSE[3]	$\hat{\sigma}^2 = \text{SSE}/(n-2)$
Total	$n-1$[4]	SST($= C_{yy}$)	

[1]Mean square = Sum of squares/d.f.
[2]1 d.f. corresponds to the regression coefficient b.
[3]Obtained by differencing (that is, total–regression).
[4]$n-1$ since we have corrected for the mean.

If $\beta = 0$, SSR will give an estimate of σ^2 on 1 d.f. while $\hat{\sigma}^2 = \text{SSE}/(n-2)$ gives us a second independent estimate of σ^2 on $n-2$ d.f. A test of the hypotheses

$$H_0 : \beta = 0$$

$$H_1 : \beta \neq 0$$

can be based on the ratio

$$F = \frac{\text{SSR}}{\hat{\sigma}^2} \quad .$$

If H_0 is true and the random experimental error ϵ_j in the linear model (6.1) is normally distributed (with mean zero and unknown variance σ^2) the ratio F has Fisher's F-distribution (section 5.6) with $\nu_1 = 1$ and $\nu_2 = n-2$ d.f.s. Here ν_1 and ν_2 are the numbers of d.f.s associated with the variance estimates on the numerator and denominator of the F-ratio.

If H_0 is not true it can be shown that the F-ratio will tend to be large (and positive), irrespective of the direction in which β differs from zero, so that we use a one-tailed (upper tail) test for testing H_1 against H_0.

The required percentage points of the F-ratio statistic may be found from Table A4.

Ex. 6.2. Test the significance of the regression for the data of example 6.1.

In addition to the quantities evaluated in example 6.1 we need

$$C_{yy} = \Sigma y^2 - \frac{1}{11}(\Sigma y)^2 = 11\ 811 - \frac{1}{11} \times 317^2 = 2675.6364$$

and SSR $= C_{xy}^2/C_{xx} = 935.6364^2/399.6364 = 2190.5299$.

ANOVA table

Source of variation	d.f.	Sum of squares	Mean square	F-ratio
Regression	1	2190.5299	2190.5299	40.64
Residual	9	485.1065	53.9007	
Total	10	2675.6344		

From Table A4 $F_{1,9}(0.05) = 5.12, F_{1,9}(0.01) = 10.56, F_{1,9}(0.001) = 22.86$. Hence F is significant at the 0.1% level and there is almost conclusive evidence that the regression coefficient differs from zero (and hence that y does depend on x).

Having established that β differs from zero we often wish to make further inferences, such as finding confidence intervals for β, or for $\alpha + \beta x$. On the basis of the model (6.1) for the observations and the assumption that the ϵ_js are independently and identically normally distributed with mean 0 and variance σ^2, the following results may be obtained

Sampling distribution of b

$$b \sim N\left(\beta; \frac{\sigma^2}{C_{xx}}\right) \quad \text{or} \quad \frac{b-\beta}{\sigma/\sqrt{C_{xx}}} \sim N(0;1) \qquad (6.12)$$

Sampling distribution of a

$$a \sim N\left(\alpha; \sigma^2 \left(\frac{1}{n} + \frac{\bar{x}^2}{C_{xx}}\right)\right) \quad \text{or} \quad \frac{a - \alpha}{\sigma \sqrt{\left(\frac{1}{n} + \frac{\bar{x}^2}{C_{xx}}\right)}} \sim N(0; 1)$$

(6.13)

Sampling distribution of $\hat{y}_j$

$$\hat{y}_j \sim N\left(\alpha + \beta x_j; \sigma^2 \left(\frac{1}{n} + \frac{(x_j - \bar{x})^2}{C_{xx}}\right)\right)$$

$$\text{or} \quad \frac{\hat{y}_j - \alpha - \beta x_j}{\sigma \sqrt{\left(\frac{1}{n} + \frac{(x_j - \bar{x})^2}{C_{xx}}\right)}} \sim N(0; 1)$$

(6.14)

If σ is known the above results may be used to carry out hypothesis tests and calculate confidence limits using critical values and confidence coefficients obtained from Table A1(b).

However, usually σ is not known and in this case must be replaced by its estimate $\hat{\sigma}$ (on $n - 2$ d.f.s) obtained from the ANOVA table. Then the statistics

$$\frac{b - \beta}{\hat{\sigma}/\sqrt{C_{xx}}}, \quad \frac{a - \alpha}{\hat{\sigma} \sqrt{\left(\frac{1}{n} + \frac{\bar{x}^2}{C_{xx}}\right)}} \quad \text{and} \quad \frac{\hat{y}_j - \alpha - \beta x_j}{\hat{\sigma} \sqrt{\left(\frac{1}{n} + \frac{(x_j - \bar{x})^2}{C_{xx}}\right)}}$$

each follow Student's t-distribution with $n - 2$ d.f.s (section 5.7). Note that the distribution for a is a particular case of the distribution for $\hat{y}_j$ with $x_j = 0$.

Ex. 6.3. For the data of example 6.1, (a) test the hypotheses $H_0 : \beta = 0$, $H_1 : \beta \neq 0$, (b) find 95% confidence limits for β, (c) find 95% confidence limits for α, (d) find 95% confidence limits for $\alpha + \beta x_j$.

(a) $H_0 : \beta = 0$

             $H_1 : \beta \neq 0$.

The test statistic is

$$t = \frac{b - 0}{\hat{\sigma}/\sqrt{C_{xx}}}$$

which under H_0 will have a t-distribution with $\nu = 11 - 2 = 9$ d.f.s. The sample value for t is

$$t = \frac{2.3412}{\dfrac{7.3417}{\sqrt{(399.6364)}}}$$

$$= 6.37(49) \ .$$

From Table A2, for a two-tailed test

$$t_9(0.025) = 2.262, \quad t_9(0.005) = 3.250, \quad t_9(0.0005) = 4.781 \ .$$

Hence t is significant at the 0.1% level and there is almost conclusive evidence that β differs from zero.

This is the same conclusion arrived at in example 6.2. Indeed the tests are exactly equivalent as may be seen by comparing the square roots of the sample F-value and the percentage points for the F-test with the corresponding values in the t-test above.

(b) 95% confidence limits for β are given by

$$b \pm t_9(0.025) \ \frac{\hat{\sigma}}{\sqrt{C_{xx}}}$$

$$= \ 2.3412 \pm 2.262 \times \frac{7.3417}{\sqrt{(399.6364)}}$$

$$= \ 1.51(05), \ 3.17(19) \ .$$

(c) 95% confidence limits for α are given by

$$a \pm t_9(0.025) \ \hat{\sigma} \ \sqrt{\left(\frac{1}{n} + \frac{\bar{x}^2}{C_{xx}}\right)}$$

$$= \ 8.1729 \pm 2.262 \times 7.3417 \ \sqrt{\left(\frac{1}{11} + \frac{8.8182^2}{399.6364}\right)}$$

$$= \ -0.7004, \ 17.0462$$

$$\doteq \ -0.70, \quad 17.05 \ .$$

It can be shown that the distributions of a and b are not independent so that in considering both α and β we should really be looking at a confidence region in two dimensions. However the method of doing this is outside the scope of the present text.

(d) The required 95% confidence limits are

$$\hat{y}_j \pm 2.262 \times 7.3417 \sqrt{\left(\frac{1}{11} + \frac{(x_j - 8.8182)^2}{399.6364} \right)}$$

If $x = 5$ for example the limits are 19.879 ± 5.9273

$$= 13.95, \quad 25.81 \ .$$

Proceeding in this way we find

x	0	5	10	15	20
Lower limit	−0.70	13.95	26.48	36.12	44.44
Upper limit	17.05	25.81	36.69	50.46	65.55

The confidence bands for $\alpha + \beta x$ are sketched in Fig. 6.5.

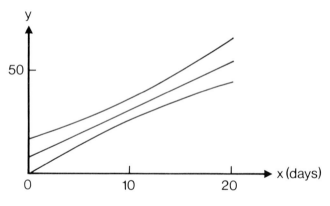

Fig. 6.5 − 95% confidence bands for $\alpha + \beta x$.

The method of least squares may be extended to fit more complex curves to experimental responses. These include polynomial curves in a single explanatory variable x of the form $y = a + b_1 x + b_2 x^2 + \ldots + b_k x^k$, and also situations in which there is more than one explanatory variable. For example if the yield y in a chemical reaction depends on two explanatory variables x_1 and x_2 (where x_1 could be temperature and x_2 pressure) we might try to fit a model of the form $y = a + b_1 x_1 + b_2 x_2$.

Computer packages are available which carry out the complex arithmetic necessary for fitting such models using the method of least squares. These programs also output the information necessary for testing the significance of the regression coefficients obtained.

6.3 CORRELATION

So far we have considered the situation in which the explanatory variable x is under the control of the experimenter in so far as he can set the value of x to any required value. A different situation arises when we draw a random sample of size n from a population, make observations x on X and y on Y, where X and Y are random variables associated with the population members, to give a set of observations $(x_1, y_1), (x_2, y_2), \ldots, (x_n, y_n)$. We then try to establish a linear relationship between x and y on the basis of these measurements. A simple example would be if we chose a random sample of 20 girls from a large class and measured their heights x m and weights y kg. If the experiment was repeated with another random sample of 20 both the x and y values would be different for the two samples.

We shall be particularly interested in the case where the variables X and Y jointly follow a *bivariate normal distribution*. This distribution has the property that the random variable X considered by itself has a normal distribution (with mean μ_x and variance σ_x^2, say) while the random variable Y considered by itself has a normal distribution (with mean μ_y and variance σ_y^2, say). The bivariate normal distribution is then defined by the parameters $\mu_x, \sigma_x^2, \mu_y, \sigma_y^2$ together with an additional parameter ρ where ρ is the *population correlation coefficient* between X and Y. The coefficient ρ is a measure of the correlation or linkage between X and Y. It can be shown that $-1 \leqslant \rho \leqslant 1$; values of ρ near $+1$ indicate a strong positive correlation between X and Y (Y increases as X increases), values of ρ near -1 indicate a strong negative correlation between X and Y (Y decreases as X increases) while if $\rho = 0$, X and Y are statistically independent.

The sample estimate of ρ is the *sample correlation coefficient r*, defined by

$$r = \frac{C_{xy}}{\sqrt{C_{xx}C_{yy}}} \tag{6.15}$$

It can be shown that r, like ρ, lies between -1 and $+1$.

If X and Y have a bivariate normal distribution it can be shown that the mean value of Y given x lies on a straight line (the line of regression of y on x). The best estimate of this line based on the sample measurements $(x_1, y_1), \ldots, (x_n, y_n)$ is

$$y = a + bx$$

where $b = C_{xy}/C_{xx}$ and $a = \bar{y} - b\bar{x}$. These are of course the same results which were obtained for the least squares estimates of α and β (equations (6.3) and (6.4)) when the x variable was controlled; hence although the two models we started from are quite different the method of obtaining the estimate of the linear relationship is the same for each.

Because X is a random variable there will also be a line of regression of x on y; from symmetry this may be estimated from

$$x = a' + b'y \qquad (6.16)$$

where $b' = C_{xy}/C_{yy}$ and $a' = \bar{x} - b'\bar{y}$.

These lines both pass through the point $(\bar{x}, \bar{y})$ but in general they will not coincide (for the exceptions see the next paragraph).

Three cases are of especial interest (a) $r = \pm 1$ (b) $r = 0$

(a) $r = \pm 1$. It can be shown that when $r = \pm 1$ the regression lines coincide and all the experimental points lie exactly on the common line of regression.

When $r = +1$, y increases as x increases and we have perfect positive correlation between the sample values of x and y (Fig. 6.6(a)).

When $r = -1$, y decreases as x increases and we have perfect negative correlation between the sample values of x and y (Fig. 6.6(b)).

(b) $r = 0$. Since $r = 0$, $C_{xy} = 0$ and the lines of regression become $y = \bar{y}$ and $x = \bar{x}$. In this case the set of values of x and y are said to be uncorrelated and the lines of regression are at right angles to each other (Fig. 6.6(c)).

Test of the null hypothesis $H_0 : \rho = 0$

The sample estimate r of ρ will have a sampling distribution. It can be shown that if $\rho = 0$ the statistic

$$t = \frac{r\sqrt{n-2}}{\sqrt{1-r^2}} \qquad (6.17)$$

has Student's t-distribution with $\nu = n-2$ d.f. Hence when r has been calculated

from $r = \dfrac{C_{xy}}{\sqrt{C_{xx}C_{yy}}}$ the statistic can be evaluated and compared with the

percentage points of t_{n-2} in the usual way.

This test is only valid for testing $H_0 : \rho = 0$. In order to test $H_0 : \rho = \rho_0 (\rho_0 \neq 0)$ a different test, which will not be discussed here, must be employed.

Ex. 6.4. An experiment was carried out on isolated guinea pig hearts to measure the rate of beating of the hearts before (x beats/min) and after (y beats/min) a specified dose of phenoxybenzamine. The results obtained are tabulated below.

x	136	140	160	183	190	196	205	210	221	221	234
y	125	150	173	219	208	227	260	243	276	262	284

Find the lines of regression of y on x and of x on y. Plot the experimental points and the lines of regression on a scatter diagram.

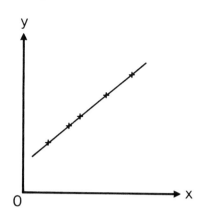

Fig. 6.6(a) – Perfect positive correlation.

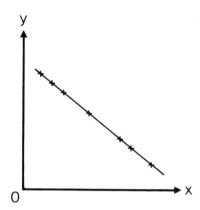

Fig. 6.6(b) – Perfect negative correlation.

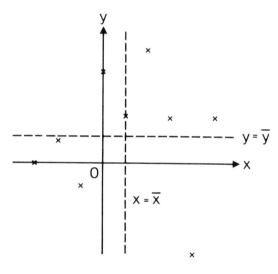

Fig. 6.6(c) – Zero correlation.

Find the coefficient of correlation between x and y and test whether it is significantly greater than zero.

For the above data we find $\Sigma x = 2096$, $\Sigma x^2 = 410264$, $\Sigma y = 2427$, $\Sigma y^2 = 562933$ and $\Sigma xy = 479453$.

Then $\bar{x} = \Sigma x/11 = 190.55$, $C_{xx} = \Sigma x^2 - (\Sigma x)^2/11 = 10880.73$

$\bar{y} = \Sigma y/11 = 220.64$, $C_{yy} = \Sigma y^2 - (\Sigma y)^2/11 = 27448.55$

$C_{xy} = \Sigma xy - (\Sigma x)(\Sigma y)/11 = 16999.18$.

Coefficient of correlation

$$r = \frac{C_{xy}}{\sqrt{C_{xx}C_{yy}}} = 0.9836 \ .$$

Regression of y on x

$$b = \frac{C_{xy}}{C_{xx}} = 1.5623, \ a = \bar{y} - b\bar{x} = -77.0566 \ .$$

Then

$$y = -77.1 + 1.56x \ .$$

Regression of x on y

$$b' = \frac{C_{xy}}{C_{yy}} = 0.6193, \ a' = \bar{x} - b'\bar{y} = 53.9030 \ .$$

Then

$$x = 53.9 + 0.62y \ .$$

Finally we wish to test

$$\left. \begin{array}{l} H_0 : \rho = 0 \\ \text{against} \qquad H_1 : \rho > 0 \end{array} \right\} \text{ one-tailed test}$$

$$t = \frac{r\sqrt{n-2}}{\sqrt{1-r^2}}$$

$$= \frac{0.9836\sqrt{9}}{\sqrt{(1-0.9836^2)}}$$

$$= 16.36 \quad \text{on 9 d.f.}$$

From Table A2, $t_9(0.001) = 4.297$.

Hence the result is significant at the 0.1% level and there is almost conclusive evidence for accepting the hypothesis that the population correlation coefficient is positive.

The data and the regression lines are plotted on a scatter diagram in Fig. 6.7.

6.4 USE OF TRANSFORMATIONS TO OBTAIN LINEARITY

Exact relationships of the form $y = a.b^x$ and $y = a.x^k$ may be reduced to linear forms by taking logarithms to give $Y = \log a + x \log b$ and $Y = \log a + kX$

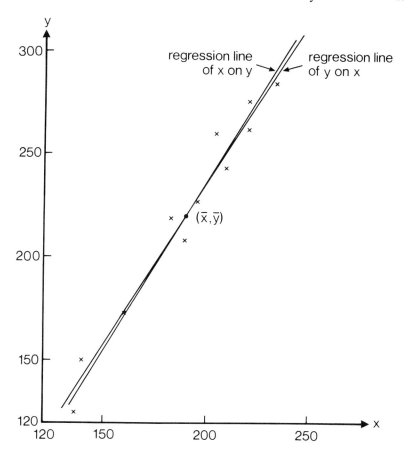

Fig. 6.7 – Regression lines and data for example 6.4.

(where $Y = \log y$ and $X = \log x$) respectively. Hence if we plot Y against x in the first case, and Y against X in the second case, straight lines will be obtained. Similar transformations may be applied when the relationships are not exact as illustrated in the following example.

Ex. 6.5. The oxygen consumption C (μl/h) of six specimens of an organism were measured using a respirometer. The organisms were then killed, dried and their weights W g obtained. The results of the experiment are tabulated below.

C	70.8	537.0	6.31	708.0	955.0	91.2
W	0.195	0.955	0.100	1.660	6.31	1.000

Find the line of regression of $y = \log_{10} C$ on $x = \log_{10} W$. Hence find a relationship between C and W and estimate the value of C when $W = 0.2$.

Taking logarithms to the base 10 we obtain

y	1.850	2.730	0.800	2.850	2.980	1.960
x	−0.710	−0.020	−1.000	0.220	0.800	0.000

The scatter diagrams of C against W and y against x are plotted in Figs 6.8(a) and 6.8(b).

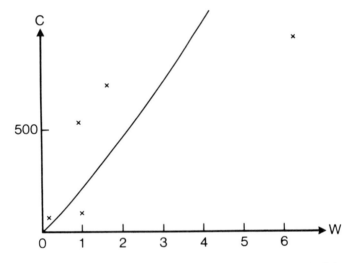

Fig. 6.8(a) – Scatter diagram of C against W for the data of example 6.5.

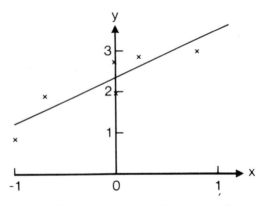

Fig. 6.8(b) – Scatter diagram of y against x for the data of example 6.5.

For the transformed data

$n = 6$ $\Sigma x = -0.71$, $\Sigma x^2 = 2.1929$, $\Sigma y = 13.17$, $\Sigma xy = 0.8429$

$$\bar{x} = -0.1183 \quad C_{xx} = 2.1929 - \frac{1}{6}(-0.71)^2 = 2.1089$$

$$\bar{y} = 2.1950 \quad C_{xy} = 0.8429 - \frac{1}{6}(-0.71)(13.17) = 2.4014 .$$

Then $b = C_{xy}/C_{xx} = 2.4014/2.1089 = 1.1387$,

and $a = \bar{y} - b\bar{x} = 2.195 - 1.1387 \times (-0.1183) = 2.3297$.

Hence the regression line of y on x has equation

$\qquad\qquad y = 2.3297 + 1.1387x$

or $\qquad\qquad \log_{10} C = 2.3297 + 1.1387 \log_{10} W$.

Thus

$\qquad\qquad C = 213.65 W^{1.1387}$.

When

$\qquad\qquad W = 0.2 \quad \hat{C} = 34.2$.

A curve fitting program (based on the method of least squares) may be used to fit a relationship $C = a_1 W^{b_1}$. This leads to

$\qquad\qquad C = 365.58 W^{0.54940}$

and the corresponding line $y = 2.5630 + 0.54940x$. Thus differing results are obtained depending on whether the sum of squares of the residuals is minimised for the points (x_i, y_i) or the points (W_i, C_i) $(i = 1, \ldots, n)$. If the underlying model for the data is

$\qquad\qquad y = \alpha + \beta x + \epsilon$

the relationship $y = 2.3297 + 1.1387x$ will be preferred, if the underlying model for the data is

$\qquad\qquad C = \alpha_1 W^{\beta_1} + \epsilon$

the relationship

$\qquad\qquad C = 365.58 W^{0.54940}$

will be preferred.

6.5 BASIC PROGRAMS FOR REGRESSION AND CORRELATION

The structure of program 6.1 is illustrated in the flow diagram below:

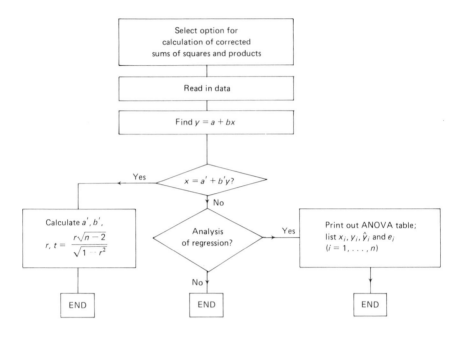

In the program C_{xx} and C_{xy} may be calculated using either of two options; in option A the means $\bar{x}$ and $\bar{y}$ are first calculated and then C_{xx} and C_{xy} are calculated using formulae (6.5a) and (6.6a) respectively; in option B, C_{xx} and C_{xy} are calculated using formulae (6.5b) and (6.6b). C_{yy} is calculated similarly in each case.

The output using both options is illustrated using the data of example 6.4. Because the x and y values are spread over wide ranges, the results obtained using both methods of calculation C_{xx} and C_{xy} are virtually identical. A set of data for which this is not the case is considered in the exercises at the end of this chapter (section C, number 3).

The regression analysis output option for the program is illustrated using the data of example 6.1 and option B. Once again virtually the same output may be obtained using option A.

Program listing and sample output

```
100 PRINT"[]":REM**CLEAR SCREEN
110 PRINT"        **PROGRAM 6.1**"
120 PRINT"  REGRESSION LINES FOR N POINTS"
130 PRINT"  X,Y READ IN FROM DATA STATEMENTS"
140 PRINT
150 PRINT"OPTION A-FORMULAE 6.5A AND 6.6A"
160 PRINT"  USED TO CALCULATE CXX AND CXY"
170 PRINT"OPTION B-FORMULAE 6.5B AND 6.6B"
180 PRINT"  USED TO CALCULATE CXX AND CXY"
190 PRINT
200 DIM X(50),Y(50)
210 PRINT"DO YOU REQUIRE OPTION A (Y/N)";
220 INPUT A$
230 READ N
240 REM**ZERO SUMS,SUMS OF SQUARES AND PRODUCTS
250 S1X=0 : S2X2=0 : S3Y=0 : S4Y2=0 : S5XY=0
260 IF A$="N" GOTO 410
270 REM**CALCULATIONS USING 6.5A AND 6.6A
280 FOR I=1 TO N
290 READ X(I),Y(I)
300 S1X=S1X+X(I)
310 S3Y=S3Y+Y(I)
320 NEXT I
330 MX=S1X/N : MY=S3Y/N
340 FOR I=1 TO N
350 DX=X(I)-MX : DY=Y(I)-MY
360 C1XX=C1XX+DX*DX
370 C2YY=C2YY+DY*DY
380 C3XY=C3XY+DX*DY
390 NEXT I
400 GOTO 530
410 REM**CALCULATIONS USING 6.5B AND 6.6B
420 FOR I=1 TO N
430 READ X(I),Y(I)
440 S1X=S1X+X(I)
450 S2X2=S2X2+X(I)*X(I)
460 S3Y=S3Y+Y(I)
470 S4Y2=S4Y2+Y(I)*Y(I)
480 S5XY=S5XY+X(I)*Y(I)
490 NEXT I
500 C1XX=S2X2-S1X*S1X/N
510 C2YY=S4Y2-S3Y*S3Y/N
520 C3XY=S5XY-S1X*S3Y/N
530 B=C3XY/C1XX
540 A=(S3Y-S1X*B)/N
550 PRINT
560 PRINT"REGRESSION LINE OF Y ON X IS:-"
570 PRINT"  Y=";A;"+";B;"*X"
580 PRINT:PRINT
590 PRINT"DO YOU REQUIRE THE LINE OF REGRESSION OF X ON Y (Y/N)";
600 INPUT A$ : IF A$="Y" THEN 1050
610 PRINT"DO YOU REQUIRE THE ANALYSIS OF REGRESSION FOR Y ON X (Y/N)";
620 INPUT A$ : IF A$="N" THEN END
630 REM**ANALYSIS OF REGRESSION
640 SR=C3XY*C3XY/C1XX
650 SE=C2YY-SR
660 MSE=SE/(N-2)
670 F=INT(100*SR/MSE+0.5)/100
680 SR=INT(100*SR+0.5)/100
690 SE=INT(100*SE+0.5)/100
```

```
700 MSE=INT(100*MSE+0.5)/100
710 C2YY=INT(100*C2YY+0.5)/100
720 PRINT"▊":REM**CLEAR SCREEN
730 PRINT"ANOVA TABLE"
740 FOR I=1 TO 11 : PRINT"-"; : NEXT I
750 PRINT
760 PRINT"SOURCE D.F. SUM OF SQ. MEAN SQ.    F"
770 FOR I=1 TO 40 : PRINT"-"; : NEXT I
780 PRINT
790 PRINT"REG.";TAB(6);" 1";TAB(10)SR;TAB(21);SR;TAB(30);F
800 PRINT"RES.";TAB(6);N-2;TAB(10);SE;TAB(21);MSE
810 FOR I=1 TO 40 : PRINT"-"; : NEXT I
820 PRINT
830 PRINT"TOT.";TAB(6);N-1;TAB(10);C2YY
840 PRINT
850 PRINT"PRESS ANY KEY TO CONTINUE"
860 GET A$ : IF A$="" THEN 860
870 PRINT"▊":REM**CLEAR SCREEN
880 T1E=0 : T2E2=0
890 PRINT" I     X(I) Y(I)      YHAT(I)     E(I)"
900 FOR I=1 TO N
910 YHAT=A+B*X(I)
915 E=Y(I)-YHAT
920 T1E=T1E+E
930 T2E2=T2E2+E*E
940 YHAT=INT(100*YHAT+0.5)/100
950 E=INT(100*E+0.5)/100
960 PRINT I;TAB(6);X(I);TAB(10);Y(I);TAB(20);YHAT;TAB(30);E
970 IF I<>INT(I/10)*10 THEN 990
980 PRINT"PRESS ANY KEY TO CONTINUE"
985 GET A$ : IF A$="" THEN 985
990 NEXT I
1000 T1E=INT(100*T1E+0.5)/100
1010 T2E2=INT(100*T2E2+0.5)/100
1020 PRINT:PRINT"SUM OF RESIDUALS=";T1E
1030 PRINT"SUM OF SQUARES OF THE RESIDUALS=";T2E2
1040 PRINT:PRINT : END
1050 REM**REGRESSION OF X ON Y AND
1060 REM**CALCULATION OF CORRELATION COEFFICIENT
1070 B1=C3XY/C2YY
1080 A1=(S1X-S3Y*B1)/N
1090 PRINT"▊":REM**CLEAR SCREEN
1100 PRINT"REGRESSION LINE OF X ON Y IS:-"
1110 PRINT" X=";A1;"+";B1;"*Y"
1120 R=C3XY/SQR(C1XX*C2YY)
1130 T=R*SQR((N-2)/(1-R*R))
1140 R=INT(10000*R+0.5)/10000
1150 T=INT(100*T+C.5)/100
1160 PRINT:PRINT"CORRELATION COEFFICIENT IS ";R
1170 PRINT:PRINT"T-STATISTIC ON ";N-2;"D.F.'S FOR"
1180 PRINT"TESTING HO:RHO=0 IS ";T
1190 PRINT:PRINT
1200 DATA 11
1210 DATA 136,125,140,150,160,173,183,219,190,208
1220 DATA 196,227,205,260,210,243,221,276
1230 DATA 221,262,234,284
1240 END
```

Data for example 6.4 (option A)

```
DO YOU REQUIRE OPTION A(Y/N) N

REGRESSION LINE OF Y ON X IS:-
  Y=-77.0566802 + 1.56232036 *X


REGRESSION LINE OF X ON Y IS:-
  X= 53.9029623 + .619310843 *Y

CORRELATION COEFFICIENT IS  .9836

T-STATISTIC ON  9 D.F.'S FOR
TESTING HO:RHO=0 IS  16.38
```

Data for example 6.4 (option B)

```
DO YOU REQUIRE OPTION A(Y/N) Y

REGRESSION LINE OF Y ON X IS:-
  Y=-77.0566807 + 1.56232037 *X


REGRESSION LINE OF X ON Y IS:-
  X= 53.9029623 + .619310843 *Y

CORRELATION COEFFICIENT IS  .9836

T-STATISTIC ON  9 D.F.'S FOR
TESTING HO:RHO=0 IS  16.38
```

Data for Example 6.1 (option B)

```
DO YOU REQUIRE OPTION A(Y/N) N

REGRESSION LINE OF Y ON X IS:-
  Y= 8.17288444 + 2.34121929 *X

ANOVA TABLE
-----------
```

SOURCE	D.F	SUM OF SQ.	MEAN SQ.	F
REG.	1	2190.53	2190.53	40.64
RES.	9	485.11	53.9	
TOT.	10	2675.64		

```
TABLE OF RESIDUALS
------------------
```

I	X(I)	Y(I)	YHAT(I)	E(I)
1	1	6	10.51	-4.51
2	1	9	10.51	-1.51
3	3	12	15.2	-3.2
4	5	18	19.88	-1.88
5	6	31	22.22	8.78
6	10	38	31.59	6.41
7	10	44	31.59	12.41
8	11	22	33.93	-11.93
9	14	37	40.95	-3.95
10	15	46	43.29	2.71
11	21	54	57.34	-3.34

```
SUM OF RESIDUALS= 0
SUM OF SQUARES OF THE RESIDUALS= 485.11
```

APPENDIX 6.1 DERIVATION OF THE LEAST SQUARES ESTIMATES OF α AND β

We have from (6.2)

$$S = \sum_{i=1}^{n} (y_i - a - bx_i)^2 \ .$$

Then $\partial S/\partial a = 0$ gives

$$-2 \sum_{i=1}^{n} (y_i - a - bx_i) = 0$$

or

$$na + b \sum_{i=1}^{n} x_i = \sum_{i=1}^{n} y_i \ .$$

Division by n leads to

$$a + b\bar{x} = \bar{y}$$

$$a = \bar{y} - b\bar{x} \ . \tag{6.4}$$

$\partial S/\partial b = 0$ gives

$$-2 \sum_{i=1}^{n} x_i(y_i - a - bx_i) = 0$$

$$a \sum_{i=1}^{n} x_i + b \sum_{i=1}^{n} x_i^2 = \sum_{i=1}^{n} x_i y_i \ .$$

Substituting $a = \bar{y} - b\bar{x}$ in this last equation gives

$$(\bar{y} - b\bar{x})n\bar{x} + b \sum_{i=1}^{n} x_i^2 = \sum_{i=1}^{n} x_i y_i$$

$$b \left(\sum_{i=1}^{n} x_i^2 - n\bar{x}^2 \right) = \sum_{i=1}^{n} x_i y_i - n\bar{x}\bar{y}$$

and using the results of Appendix 6.2

$$b = \frac{C_{xy}}{C_{xx}} \ . \tag{6.3}$$

It can be shown that the values (6.4) and (6.3) for a and b lead to a minimum (rather than a maximum) value for the residual sum of squares.

APPENDIX 6.2 DERIVATION OF THE ALTERNATIVE FORMS FOR C_{xx} AND C_{xy}

From (6.6a)

$$C_{xy} = \sum_{i=1}^{n} (x_i - \bar{x})(y_i - \bar{y})$$

$$= \sum_{i=1}^{n} x_i y_i - \bar{x} \sum_{i=1}^{n} y_i - \bar{y} \sum_{i=1}^{n} x_i + n\bar{x}\,\bar{y}$$

$$= \sum_{i=1}^{n} x_i y_i - n\bar{x}\,\bar{y} - n\bar{x}\,\bar{y} + n\bar{x}\,\bar{y}$$

$$= \sum_{i=1}^{n} x_i y_i - n\bar{x}\,\bar{y}$$

$$= \sum_{i=1}^{n} x_i y_i - n\,\frac{1}{n}\left(\sum_{i=1}^{n} x_i\right)\frac{1}{n}\left(\sum_{i=1}^{n} y_i\right)$$

$$= \sum_{i=1}^{n} x_i y_i - \frac{1}{n}\left(\sum_{i=1}^{n} x_i\right)\left(\sum_{i=1}^{n} y_i\right). \tag{6.6b}$$

Putting $y_i = x_i$ in (6.6a) and (6.6b) gives

$$C_{xx} = \sum_{i=1}^{n} (x_i - \bar{x})^2 = \sum_{i=1}^{n} x_i^2 - \frac{1}{n}\left(\sum_{i=1}^{n} x_i\right)^2 \tag{6.5b}$$

Similarly putting $x_i = y_i$ in (6.6a) and (6.6b) gives

$$C_{yy} = \sum_{i=1}^{n} (y_i - \bar{y})^2 = \sum_{i=1}^{n} y_i^2 - \frac{1}{n}\left(\sum_{i=1}^{n} y_i\right)^2. \tag{6.7}$$

APPENDIX 6.3 ANALYSIS OF REGRESSION

$$C_{yy} = \sum_{i=1}^{n} (y_i - \bar{y})^2 = \sum_{i=1}^{n} (y_i - \hat{y}_i + \hat{y}_i - \bar{y})^2,$$

subtracting and adding $\hat{y}_i$

$$= \sum_{i=1}^{n} (y_i - \hat{y}_i)^2 + 2T + \sum_{i=1}^{n} (\hat{y}_i - \bar{y})^2$$

where

$$T = \sum_{i=1}^{n} (y_i - \hat{y}_i)(\hat{y}_i - \bar{y}) = \sum_{i=1}^{n} (y_i - a - bx_i)(a + bx_i - \bar{y})$$

$$= \sum_{i=1}^{n} (y_i - \bar{y} - b(x_i - \bar{x}))(b(x_i - \bar{x})) \text{ using } a = \bar{y} - b\bar{x}$$

$$= b \sum_{i=1}^{n} (x_i - \bar{x})(y_i - \bar{y}) - b^2 \sum_{i=1}^{n} (x_i - \bar{x})^2$$

$$= bC_{xy} - b^2 C_{xx}$$

$$= C_{xy}^2/C_{xx} - C_{xy}^2/C_{xx} \quad \text{using } b = C_{xy}/C_{xx}$$

$$= 0 .$$

Then $$C_{yy} = \sum_{i=1}^{n} (y_i - \hat{y}_i)^2 + \sum_{i=1}^{n} (\hat{y}_i - \bar{y})^2 .$$ (6.9)

APPENDIX 6.4 USE OF CODING

Let $$\left. \begin{array}{l} x_i = cu_i + \alpha \\ y_i = dv_i + \beta \end{array} \right\} \quad i = 1, 2, \ldots, n$$

Then $$\bar{x} = c\bar{u} + \alpha$$

and $$\bar{y} = d\bar{v} + \beta,$$

so that $$x_i - \bar{x} = c(u_i - \bar{u})$$

and $$y_i - \bar{y} = d(v_i - \bar{v}) .$$

From (6.5a)

$$C_{xx} = \sum_{i=1}^{n} (x_i - \bar{x})^2$$

$$= c^2 C_{uu}$$

and from (6.6a)

$$C_{xy} = \sum_{i=1}^{n} (x_i - \bar{x})(y_i - \bar{y})$$

$$= cd\, C_{uv} .$$

Then

$$b = \frac{C_{xy}}{C_{xx}} \quad \text{from (6.3)}$$

$$= \frac{dC_{uv}}{cC_{uu}}$$

and

$$a = \bar{y} - b\bar{x} \quad \text{from (6.4)}$$

$$= (d\bar{v} + \beta) - (c\bar{u} + \alpha)b.$$

EXERCISES

Section A

1. In a certain chemical process it is known that the temperature (x °C) affects the amount of a given impurity A (y%). Experimentally determined values of y at given values of x are tabulated below.

x	80	82	84	86	88	90
y	0.05	0.11	0.20	0.25	0.33	0.38

Illustrate the data on a scatter diagram.
   Find the equation of the line of regression of y on x and draw the line on the scatter diagram. Test whether the regression is significant.
   Estimate the value of y (a) when $x = 85$, (b) when $x = 92$.

2. To test the alcohol method for the quantitative analysis of calcium oxide (CaO) in the presence of magnesium oxide (MgO) known quantities x mg of CaO were added to each of ten samples of calcium-free mixture containing large amounts of MgO and analysed for its CaO content. The results of the analyses (y mg of CaO) are tabulated below.

x	4.0	8.0	12.5	16.0	20.0	25.0	31.0	36.0	40.0	40.0
y	3.7	7.8	12.1	15.6	19.8	24.5	31.1	35.5	39.4	39.5

Illustrate the experimental points on a scatter diagram. Find the line of regression of y on x and draw it on the same diagram. Assuming the model $y = \alpha + \beta x + \epsilon$, find 95% confidence limits for β.

3. The national consumption of a particular drug over an eight-year period is tabulated below.

Year	1	2	3	4	5	6	7	8
Consumption (10^6 grains)	9.1	9.3	9.5	9.7	10.0	10.4	10.8	11.2

Use the method of least squares to fit a staight line to the data. Illustrate the line and points on the same diagram. Comment on the fit.

Estimate the expected consumption of the drug in year 9 and find 95% confidence limits for this expected consumption.

4. The blood pressures and ages of a sample of eight men are tabulated below.

Age (x)	38	42	45	49	55	56	60	63
Blood pressure (y)	115	120	128	143	150	147	155	149

Find the line of regression of y on x.

Plot the experimental points and draw the fitted line on the same graph.

Find the coefficient of correlation between x and y and test whether it is significantly greater than zero.

5. A biologist is studying animal growth under a special drug and measures x the initial weight and y the increase in weight for six animals. The results are tabulated below.

x	1.58	1.80	1.93	1.94	2.08	2.25
y	0.51	0.44	0.68	0.91	1.20	1.13

Find the lines of regression of y on x and of x on y. Find the coefficient of correlation between x and y and test whether it is significant.

6. The number of bacteria per unit volume found in a tillage after x hours is given in the table below:

Number of hours x	0	1	2	3	4	5	6	7
Number of bacteria y	47	64	81	107	151	209	298	841

The relationship between x and y is known to be the form:

$$\ln y = \alpha + \beta x + \epsilon$$

After transforming the appropriate variable use the method of least squares to estimate α and β. Plot the points on a graph and superimpose the calculated line.

Use the regression line to estimate the number of bacteria per unit volume to be expected after eight hours.

Put 95% confidence limits on $\alpha + \beta x$ for $x = 0, 1, 2, \ldots, 7$.

Section B

1. The following table gives the observed values of y corresponding to the given values of x

x	0.0	0.2	0.4	0.6	0.8	1.0	1.2
y	-1.85	-1.20	-0.55	0.15	0.80	1.35	2.00

Find the least squares regression line of y on x.

2. The drag D on a body being pulled at a velocity v through a fluid is measured for various values of v and the results are are tabulated below

v	3	4	5	6	7	8	9
D	2	6	7	11	12	18	19

Use least squares to fit a relationship of the form

$$D = a + bv^2$$

to the above data. Plot a scatter diagram of D against v, and plot the fitted relationship on the diagram.

3. The table below shows the annual average daily vehicle mileages from Mondays to Fridays on trunk and classified roads in urban areas for eight consecutive years. The mileage in each case is expressed as a percentage of the mileage in year five

Year	1	2	3	4	5	6	7	8
Mileage	71	72	86	95	100	106	113	125

Fit a straight line to the data and illustrate the line and the data on the same graph.

Calculate 95% confidence limits for the annual rate of increase in the mileage.

Discuss briefly any assumptions made about the data in order to calculate these limits.

Estimate the mileage in year nine as a percentage of the mileage in year five.

4. In an experiment to measure the stiffness of a spring the length y of the spring under different loads x was measured giving

x (kg)	0	0.5	1.0	1.5	2.0
y (cm)	20.0	22.2	25.4	26.7	29.8

Find the regression line of y on x and 95% confidence limits for the extension per kilogram

5. The following measurements of the specific heat of a chemical were made in an investigation of the variation of specific heat with temperature

Temperature ($T\,^{\circ}$C)	0	10	20	30	40
Specific heat (S)	0.51	0.55	0.57	0.59	0.63

Find the least squares estimates of the parameters in the linear relationship $S = a + bT$. Find 95% confidence limits for β, the true rate of increase of S with T.

6. The table below shows the number of cars per 1000 persons (y) and the gross national product per 1000 persons (x) in million units of account for eight countries in the same year

Country	1	2	3	4	5	6	7	8
x	1.70	1.90	2.20	2.20	2.60	2.70	2.90	3.00
y	186	152	198	213	209	234	245	219

Find the line of regression of y upon x, the line of regression of x upon y and the coefficient of linear correlation between x and y. Is this coefficient signficantly greater than zero?
Comment on your conclusions.

7. In an experiment it is thought that $y = $ (conductivity)^2 is a linear function of time t at a fixed temperature. In an experiment carried out at a temperature of 28.5 $^{\circ}$C the following results are obtained

t (min)	1	3	5	7	9	11	13	15
conductivity (mhos $\times 10^3$)	1.304	1.597	1.746	1.962	2.074	2.145	2.247	2.335

Find the least squares estimate of the line $y = a + bt$.

Find 95% confidence limits for β, the population value of b, stating any assumptions made about the data.

8. A standard curve for calcium determination using an autoanalyser is con-
    structed by measuring the absorbance y for different concentrations x
    (mmol/l) of Ca^{2+} leading to the following values

x	0.625	1.25	2.5	3.75	5.0
y	22.0	33.5	51.5	59.75	75.0

(a) Plot the points on a scatter diagram and fit a straight line by eye.
(b) Fit a straight line using the method of least squares.
(c) Compare the sums of squares of the residuals for the lines in (a) and (b).
(d) Test to see whether the regression is significant.
(e) Use a t-test to test the hypotheses $H_0 : \beta = 0$, $H_1 : \beta > 0$, where β is the
     gradient of the line.
(f) Find 99% confidence limits for (i) the gradient of the line, (ii) the
     intercept of the line on the y-axis, (iii) the expected value of y when
     $x = 1$.

Section C†

1. Modify program 6.1 to input n and the data points from the video screen,
    to test the input data and allow corrections to be made to the data before
    proceeding to the calculation of the regression line.
2. Verify that the output from program 6.1 for the data of example 6.1 is
    nearly identical for both options for calculating C_{xx} and C_{xy}.
3. Apply program 6.1 with option A to the following data set, where x is a
    controlled variable.

x	1.00002	1.00004	1.00006	1.00008	1.00010	1.00012	1.00014	1.00016	1.00018	1.00020
y	3.50004	3.50007	3.50004	3.50011	3.50014	3.50024	3.50027	3.50025	3.50029	3.50028

Apply option B to the same set of data and comment on the results.

4. Write a program to calculate the line of regression of y on x using coding and
    apply the program to the data of example 3.
5. Modify program 6.1 to calculate confidence limits for α, β and $\alpha + \beta x$,
    where the required percentage points of the t-statistic are read into the
    program.

†Check the operation of the programs developed in this section on data from the examples
of sections A and B, or worked examples in the text.

6. Write a program to fit curves of the form (i) $y = a \cdot b^x$, (ii) $y = a \cdot x^b$, to given sets of data, by first transforming the data and then fitting a straight line to the transformed data.

7. Explain the discrepancy between the t-value obtained in the solution to example 6.4 in the text and the corresponding value obtained in the output from program 6.1.

Suggestions for further reading

CHAPTER 2

(a) Descriptive Statistics
Tukey, J. W. (1977) *Exploratory Data Analysis*. Addison-Wesley.
(b) Programming in BASIC
Alcock, D. (1977) *Illustrating Basic*. Cambridge University Press.

CHAPTER 3

(a) Probability Theory
Mood, A. M., Graybill, F. A., and Boes, D. C. (1974) *Introduction to the Theory of Statistics* (Chapter I). McGraw-Hill.
(b) Bayesian Probability
Lindley, D. V. (1970) *Introduction to Probability and Statistics from a Bayesian Viewpoint*. Part 1 *Probability*. Cambridge University Press.
(c) Random Number Generation
Oldknow, A., and Smith, D. (1983) *Learning Mathematics with Micros*. Ellis Horwood Limited.

CHAPTER 4

Discrete and Continuous Distributions
Mood, A. M., Graybill, F. A., and Boes, D. C. (1974) *Introduction to the Theory of Statistics* (Chapter III). McGraw-Hill.

CHAPTER 5

(a) Experimental design and analysis
  (i) Stoodley, K. D. C., Lewis, T., and Stainton, C. L. S. (1980) *Applied Statistical Techniques* (Chapter 3). Ellis Horwood Limited.
  (ii) Montgomery, D. C. (1984) *Design and Analysis of Experiments*. John Wiley & Sons Limited.

(b) Non-parametric methods
  (i)   Stoodley, K. D. C., Lewis, T., and Stainton, C. L. S. (1980) *Applied Statistical Techniques* (Chapter 4). Ellis Horwood Limited.
  (ii) Siegel, S. (1956) *Non-parametric Statistics for the Behavioral Sciences.* McGraw-Hill.
(c) Bayesian inference
  Lindley, D. V. (1970) *Introduction to Probability and Statistics from a Bayesian Viewpoint.* Part 2 *Inference,* Cambridge University Press.
(d) Theory and comparison of inference techniques
  Barnett, V. (1973) *Comparative Statistical Inference.* John Wiley & Sons Limited.

CHAPTER 6

  Regression and Correlation
  (i)   Stoodley, K. D. C., Lewis, T., and Stainton, C. L. S. (1980) *Applied Statistical Techniques* (Chapter 2). Ellis Horwood Limited.
  (ii) Draper, N. R., and Smith, H. (1981) *Applied Regression Analysis.* John Wiley & Sons Limited.

Appendix:
Statistical Tables

Appendix:

Table A1(a) — Area under the standard normal curve

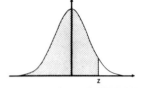

z	0.00	0.01	0.02	0.03	0.04	0.05	0.06	0.07	0.08	0.09
0.0	0.5000	0.5040	0.5080	0.5120	0.5160	0.5199	0.5239	0.5279	0.5319	0.5359
0.1	0.5398	0.5438	0.5478	0.5517	0.5557	0.5596	0.5636	0.5675	0.5714	0.5753
0.2	0.5793	0.5832	0.5871	0.5910	0.5948	0.5987	0.6026	0.6064	0.6103	0.6141
0.3	0.6179	0.6217	0.6255	0.6293	0.6331	0.6368	0.6406	0.6443	0.6480	0.6517
0.4	0.6554	0.6591	0.6628	0.6664	0.6700	0.6736	0.6772	0.6808	0.6844	0.6879
0.5	0.6915	0.6950	0.6985	0.7019	0.7054	0.7088	0.7123	0.7157	0.7190	0.7224
0.6	0.7257	0.7291	0.7324	0.7357	0.7389	0.7422	0.7454	0.7486	0.7517	0.7549
0.7	0.7580	0.7611	0.7642	0.7673	0.7704	0.7734	0.7764	0.7794	0.7823	0.7852
0.8	0.7881	0.7910	0.7939	0.7967	0.7995	0.8023	0.8051	0.8078	0.8106	0.8133
0.9	0.8159	0.8186	0.8212	0.8238	0.8264	0.8289	0.8315	0.8340	0.8365	0.8389
1.0	0.8413	0.8438	0.8461	0.8485	0.8508	0.8531	0.8554	0.8577	0.8599	0.8621
1.1	0.8643	0.8665	0.8686	0.8708	0.8729	0.8749	0.8770	0.8790	0.8810	0.8830
1.2	0.8849	0.8869	0.8888	0.8907	0.8925	0.8944	0.8962	0.8980	0.8997	0.9015
1.3	0.9032	0.9049	0.9066	0.9082	0.9099	0.9115	0.9131	0.9147	0.9162	0.9177
1.4	0.9192	0.9207	0.9222	0.9236	0.9251	0.9265	0.9279	0.9292	0.9306	0.9319
1.5	0.9332	0.9345	0.9357	0.9370	0.9382	0.9394	0.9406	0.9418	0.9429	0.9441
1.6	0.9452	0.9463	0.9474	0.9484	0.9495	0.9505	0.9515	0.9525	0.9535	0.9545
1.7	0.9554	0.9564	0.9573	0.9582	0.9591	0.9599	0.9608	0.9616	0.9625	0.9633
1.8	0.9641	0.9649	0.9656	0.9664	0.9671	0.9678	0.9686	0.9693	0.9699	0.9706
1.9	0.9713	0.9719	0.9726	0.9732	0.9738	0.9744	0.9750	0.9756	0.9761	0.9767
2.0	0.9772	0.9778	0.9783	0.9788	0.9793	0.9798	0.9803	0.9808	0.9812	0.9817
2.1	0.9821	0.9826	0.9830	0.9834	0.9838	0.9842	0.9846	0.9850	0.9854	0.9857
2.2	0.9861	0.9864	0.9868	0.9871	0.9875	0.9878	0.9881	0.9884	0.9887	0.9890
2.3	0.9893	0.9896	0.9898	0.9901	0.9904	0.9906	0.9909	0.9911	0.9913	0.9916
2.4	0.9918	0.9920	0.9922	0.9925	0.9927	0.9929	0.9931	0.9932	0.9934	0.9936
2.5	0.9938	0.9940	0.9941	0.9943	0.9945	0.9946	0.9948	0.9949	0.9951	0.9952
2.6	0.9953	0.9955	0.9956	0.9957	0.9959	0.9960	0.9961	0.9962	0.9963	0.9964
2.7	0.9965	0.9966	0.9967	0.9968	0.9969	0.9970	0.9971	0.9972	0.9973	0.9974
2.8	0.9974	0.9975	0.9976	0.9977	0.9977	0.9978	0.9979	0.9979	0.9980	0.9981
2.9	0.9981	0.9982	0.9982	0.9983	0.9984	0.9984	0.9985	0.9985	0.9986	0.9986
3.0	0.9987	0.9987	0.9987	0.9988	0.9988	0.9989	0.9989	0.9989	0.9990	0.9990
3.1	0.9990	0.9991	0.9991	0.9991	0.9992	0.9992	0.9992	0.9992	0.9993	0.9993
3.2	0.9993	0.9993	0.9994	0.9994	0.9994	0.9994	0.9994	0.9995	0.9995	0.9995
3.3	0.9995	0.9995	0.9995	0.9996	0.9996	0.9996	0.9996	0.9996	0.9996	0.9997
3.4	0.9997	0.9997	0.9997	0.9997	0.9997	0.9997	0.9997	0.9997	0.9997	0.9998

z	3.5	3.6	3.7	3.8	3.9	4.0
Area	0.99977	0.99984	0.99989	0.99993	0.99995	0.99997

Table A1(b) — Percentage points of the standard normal distribution

$z(0.05)$	$z(0.025)$	$z(0.01)$	$z(0.005)$	$z(0.001)$	$z(0.0005)$
1.645	1.96	2.33	2.58	3.09	3.29

Table A2 – Percentage points of the *t*-distribution

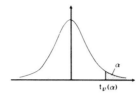

α ν	0.100	0.050	0.025	0.010	0.005	0.0025	0.001	0.0005
1	3.078	6.314	12.706	31.821	63.657	127.32	318.31	636.62
2	1.886	2.920	4.303	6.965	9.925	14.089	22.327	31.598
3	1.638	2.353	3.182	4.541	5.841	7.453	10.214	12.924
4	1.533	2.132	2.776	3.747	4.604	5.598	7.173	8.610
5	1.476	2.015	2.571	3.365	4.032	4.773	5.893	6.869
6	1.440	1.943	2.447	3.143	3.707	4.317	5.208	5.959
7	1.415	1.895	2.365	2.998	3.499	4.029	4.785	5.408
8	1.397	1.860	2.306	2.896	3.355	3.833	4.501	5.041
9	1.383	1.833	2.262	2.821	3.250	3.690	4.297	4.781
10	1.372	1.812	2.228	2.764	3.169	3.581	4.144	4.587
11	1.363	1.796	2.201	2.718	3.106	3.497	4.025	4.437
12	1.356	1.782	2.179	2.681	3.055	3.428	3.930	4.318
13	1.350	1.771	2.160	2.650	3.012	3.372	3.852	4.221
14	1.345	1.761	2.145	2.624	2.977	3.326	3.787	4.140
15	1.341	1.753	2.131	2.602	2.947	3.286	3.733	4.073
16	1.337	1.746	2.120	2.583	2.921	3.252	3.686	4.015
17	1.333	1.740	2.110	2.567	2.898	3.222	3.646	3.965
18	1.330	1.734	2.101	2.552	2.878	3.197	3.610	3.922
19	1.328	1.729	2.093	2.539	2.861	3.174	3.579	3.883
20	1.325	1.725	2.086	2.528	2.845	3.153	3.552	3.850
21	1.323	1.721	2.080	2.518	2.831	3.135	3.527	3.819
22	1.321	1.717	2.074	2.508	2.819	3.119	3.505	3.792
23	1.319	1.714	2.069	2.500	2.807	3.104	3.485	3.767
24	1.318	1.711	2.064	2.492	2.797	3.091	3.467	3.745
25	1.316	1.708	2.060	2.485	2.787	3.078	3.450	3.725
26	1.315	1.706	2.056	2.479	2.779	3.067	3.435	3.707
27	1.314	1.703	2.052	2.473	2.771	3.057	3.421	3.690
28	1.313	1.701	2.048	2.467	2.763	3.047	3.408	3.674
29	1.311	1.699	2.045	2.462	2.756	3.038	3.396	3.659
30	1.310	1.697	2.042	2.457	2.750	3.030	3.385	3.646
40	1.303	1.684	2.021	2.423	2.704	2.971	3.307	3.551
60	1.296	1.671	2.000	2.390	2.660	2.915	3.232	3.460
120	1.289	1.658	1.980	2.358	2.617	2.860	3.160	3.373
∞	1.282	1.645	1.960	2.326	2.576	2.807	3.090	3.291

Appendix:

Table A3 — Percentage points of the χ^2-distribution

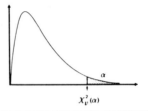

$\chi^2_\nu(\alpha)$

ν \ α	0.995	0.990	0.975	0.950
1	3.93×10^{-3}	1.57×10^{-4}	9.82×10^{-4}	3.93×10^{-3}
2	0.0100	0.0201	0.0506	0.103
3	0.0717	0.115	0.216	0.352
4	0.207	0.297	0.484	0.711
5	0.412	0.554	0.831	1.15
6	0.676	0.872	1.24	1.64
7	0.989	1.24	1.69	2.17
8	1.34	1.65	2.18	2.73
9	1.73	2.09	2.70	3.33
10	2.16	2.56	3.25	3.94
11	2.60	3.05	3.82	4.57
12	3.07	3.57	4.40	5.23
13	3.57	4.11	5.01	5.89
14	4.07	4.66	5.63	6.57
15	4.60	5.23	6.26	7.26
16	5.14	5.81	6.91	7.96
17	5.70	6.41	7.56	8.67
18	6.26	7.01	8.23	9.39
19	6.84	7.63	8.91	10.12
20	7.43	8.26	9.59	10.85
21	8.03	8.90	10.28	11.59
22	8.64	9.54	10.98	12.34
23	9.26	10.20	11.69	13.09
24	9.89	10.86	12.40	13.85
25	10.52	11.52	13.12	14.61
26	11.16	12.20	13.84	15.38
27	11.81	12.88	14.57	16.15
28	12.46	13.56	15.31	16.93
29	13.12	14.26	16.05	17.71
30	13.79	14.95	16.79	18.49
40	20.71	22.16	24.43	26.51
50	27.99	29.71	32.36	34.76
60	35.53	37.48	40.48	43.19
70	43.28	45.44	48.76	51.74
80	51.17	53.54	57.15	60.39
90	59.20	61.75	65.65	69.13
100	67.33	70.06	74.22	77.93

Table A3 (continued)

α \ ν	0.050	0.025	0.010	0.005	0.001
1	3.84	5.02	6.63	7.88	10.83
2	5.99	7.38	9.21	10.60	13.82
3	7.81	9.35	11.34	12.84	16.27
4	9.49	11.14	13.28	14.86	18.47
5	11.07	12.83	15.09	16.75	20.52
6	12.59	14.45	16.81	18.55	22.46
7	14.07	16.01	18.48	20.28	24.32
8	15.51	17.53	20.09	21.95	26.12
9	16.92	19.02	21.67	23.59	27.88
10	18.31	20.48	23.21	25.19	29.59
11	19.68	21.92	24.73	26.76	31.26
12	21.03	23.34	26.22	28.30	32.91
13	22.36	24.74	27.69	29.82	34.53
14	23.68	26.12	29.14	31.32	36.12
15	25.00	27.49	30.58	32.80	37.70
16	26.30	28.85	32.00	34.27	39.25
17	27.59	30.19	33.41	35.72	40.79
18	28.87	31.53	34.81	37.16	42.31
19	30.14	32.85	36.19	38.58	43.82
20	31.41	34.17	37.57	40.00	45.31
21	32.67	35.48	38.93	41.40	46.80
22	33.92	36.78	40.29	42.80	48.27
23	35.17	38.08	41.64	44.18	49.73
24	36.42	39.36	42.98	45.56	51.18
25	37.65	40.65	44.31	46.93	52.62
26	38.89	41.92	45.64	48.29	54.05
27	40.11	43.19	46.96	49.64	55.48
28	41.34	44.46	48.28	50.99	56.89
29	42.56	45.72	49.59	52.34	58.30
30	43.77	46.98	50.89	53.67	59.70
40	55.76	59.34	63.69	66.77	73.40
50	67.50	71.42	76.15	79.49	86.66
60	79.08	83.30	88.38	91.95	99.61
70	90.53	95.02	100.42	104.22	112.32
80	101.88	106.63	112.33	116.32	124.84
90	113.14	118.14	124.12	128.30	137.21
100	124.34	129.56	135.81	140.17	149.45

Appendix:

Table A4 – Upper percentage points of the F-distribution

(a) $F_{\nu_1, \nu_2}(0.05)$

ν_2 \ ν_1	1	2	3	4	5	6	7	8	9	10	12	15	20	24	30	40	60	120	∞
1	161.4	199.5	215.7	224.6	230.2	234.0	236.8	238.9	240.5	241.9	243.9	245.9	248.0	249.1	250.1	251.1	252.2	253.3	254.3
2	18.51	19.00	19.16	19.25	19.30	19.33	19.35	19.37	19.38	19.40	19.41	19.43	19.45	19.45	19.46	19.47	19.48	19.49	19.50
3	10.13	9.55	9.28	9.12	9.01	8.94	8.89	8.85	8.81	8.79	8.74	8.70	8.66	8.64	8.62	8.59	8.57	8.55	8.53
4	7.71	6.94	6.59	6.39	6.26	6.16	6.09	6.04	6.00	5.96	5.91	5.86	5.80	5.77	5.75	5.72	5.69	5.66	5.63
5	6.61	5.79	5.41	5.19	5.05	4.95	4.88	4.82	4.77	4.74	4.68	4.62	4.56	4.53	4.50	4.46	4.43	4.40	4.36
6	5.99	5.14	4.76	4.53	4.39	4.28	4.21	4.15	4.10	4.06	4.00	3.94	3.87	3.84	3.81	3.77	3.74	3.70	3.67
7	5.59	4.74	4.35	4.12	3.97	3.87	3.79	3.73	3.68	3.64	3.57	3.51	3.44	3.41	3.38	3.34	3.30	3.27	3.23
8	5.32	4.46	4.07	3.84	3.69	3.58	3.50	3.44	3.39	3.35	3.28	3.22	3.15	3.12	3.08	3.04	3.01	2.97	2.93
9	5.12	4.26	3.86	3.63	3.48	3.37	3.29	3.23	3.18	3.14	3.07	3.01	2.94	2.90	2.86	2.83	2.79	2.75	2.71
10	4.96	4.10	3.71	3.48	3.33	3.22	3.14	3.07	3.02	2.98	2.91	2.85	2.77	2.74	2.70	2.66	2.62	2.58	2.54
11	4.84	3.98	3.59	3.36	3.20	3.09	3.01	2.95	2.90	2.85	2.79	2.72	2.65	2.61	2.57	2.53	2.49	2.45	2.40
12	4.75	3.89	3.49	3.26	3.11	3.00	2.91	2.85	2.80	2.75	2.69	2.62	2.54	2.51	2.47	2.43	2.38	2.34	2.30
13	4.67	3.81	3.41	3.18	3.03	2.92	2.83	2.77	2.71	2.67	2.60	2.53	2.46	2.42	2.38	2.34	2.30	2.25	2.21
14	4.60	3.74	3.34	3.11	2.96	2.85	2.76	2.70	2.65	2.60	2.53	2.46	2.39	2.35	2.31	2.27	2.22	2.18	2.13
15	4.54	3.68	3.29	3.06	2.90	2.79	2.71	2.64	2.59	2.54	2.48	2.40	2.33	2.29	2.25	2.20	2.16	2.11	2.07
16	4.49	3.63	3.24	3.01	2.85	2.74	2.66	2.59	2.54	2.49	2.42	2.35	2.28	2.24	2.19	2.15	2.11	2.06	2.01
17	4.45	3.59	3.20	2.96	2.81	2.70	2.61	2.55	2.49	2.45	2.38	2.31	2.23	2.19	2.15	2.10	2.06	2.01	1.96
18	4.41	3.55	3.16	2.93	2.77	2.66	2.58	2.51	2.46	2.41	2.34	2.27	2.19	2.15	2.11	2.06	2.02	1.97	1.92
19	4.38	3.52	3.13	2.90	2.74	2.63	2.54	2.48	2.42	2.38	2.31	2.23	2.16	2.11	2.07	2.03	1.98	1.93	1.88
20	4.35	3.49	3.10	2.87	2.71	2.60	2.51	2.45	2.39	2.35	2.28	2.20	2.12	2.08	2.04	1.99	1.95	1.90	1.84
21	4.32	3.47	3.07	2.84	2.68	2.57	2.49	2.42	2.37	2.32	2.25	2.18	2.10	2.05	2.01	1.96	1.92	1.87	1.81
22	4.30	3.44	3.05	2.82	2.66	2.55	2.46	2.40	2.34	2.30	2.23	2.15	2.07	2.03	1.98	1.94	1.89	1.84	1.78
23	4.28	3.42	3.03	2.80	2.64	2.53	2.44	2.37	2.32	2.27	2.20	2.13	2.05	2.01	1.96	1.91	1.86	1.81	1.76
24	4.26	3.40	3.01	2.78	2.62	2.51	2.42	2.36	2.30	2.25	2.18	2.11	2.03	1.98	1.94	1.89	1.84	1.79	1.73
25	4.24	3.39	2.99	2.76	2.60	2.49	2.40	2.34	2.28	2.24	2.16	2.09	2.01	1.96	1.92	1.87	1.82	1.77	1.71
26	4.23	3.37	2.98	2.74	2.59	2.47	2.39	2.32	2.27	2.22	2.15	2.07	1.99	1.95	1.90	1.85	1.80	1.75	1.69
27	4.21	3.35	2.96	2.73	2.57	2.46	2.37	2.31	2.25	2.20	2.13	2.06	1.97	1.93	1.88	1.84	1.79	1.73	1.67
28	4.20	3.34	2.95	2.71	2.56	2.45	2.36	2.29	2.24	2.19	2.12	2.04	1.96	1.91	1.87	1.82	1.77	1.71	1.65
29	4.18	3.33	2.93	2.70	2.55	2.43	2.35	2.28	2.22	2.18	2.10	2.03	1.94	1.90	1.85	1.81	1.75	1.70	1.64
30	4.17	3.32	2.92	2.69	2.53	2.42	2.33	2.27	2.21	2.16	2.09	2.01	1.93	1.89	1.84	1.79	1.74	1.68	1.62
40	4.08	3.23	2.84	2.61	2.45	2.34	2.25	2.18	2.12	2.08	2.00	1.92	1.84	1.79	1.74	1.69	1.64	1.58	1.51
60	4.00	3.15	2.76	2.53	2.37	2.25	2.17	2.10	2.04	1.99	1.92	1.84	1.75	1.70	1.65	1.59	1.53	1.47	1.39
120	3.92	3.07	2.68	2.45	2.29	2.17	2.09	2.02	1.96	1.91	1.83	1.75	1.66	1.61	1.55	1.50	1.43	1.35	1.25
∞	3.84	3.00	2.60	2.37	2.21	2.10	2.01	1.94	1.88	1.83	1.75	1.67	1.57	1.52	1.46	1.39	1.32	1.22	1.00

Table A4 (Continued)

(b) F_{ν_1,ν_2} (0.025)

$\nu_2 \backslash \nu_1$	1	2	3	4	5	6	7	8	9	10	12	15	20	24	30	40	60	120	∞
1	647.8	799.5	864.2	899.6	921.8	937.1	948.2	956.7	963.3	968.6	976.7	984.9	993.1	997.2	1001	1006	1010	1014	1018
2	38.51	39.00	39.17	39.25	39.30	39.33	39.36	39.37	39.39	39.40	39.41	39.43	39.45	39.46	39.46	39.47	39.48	39.49	39.50
3	17.44	16.04	15.44	15.10	14.88	14.73	14.62	14.54	14.47	14.42	14.34	14.25	14.17	14.12	14.08	14.04	13.99	13.95	13.90
4	12.22	10.65	9.98	9.60	9.36	9.20	9.07	8.98	8.90	8.84	8.75	8.66	8.56	8.51	8.46	8.41	8.36	8.31	8.26
5	10.01	8.43	7.76	7.39	7.15	6.98	6.85	6.76	6.68	6.62	6.52	6.43	6.33	6.28	6.23	6.18	6.12	6.07	6.02
6	8.81	7.26	6.60	6.23	5.99	5.82	5.70	5.60	5.52	5.46	5.37	5.27	5.17	5.12	5.07	5.01	4.96	4.90	4.85
7	8.07	6.54	5.89	5.52	5.29	5.12	4.99	4.90	4.82	4.76	4.67	4.57	4.47	4.42	4.36	4.31	4.25	4.20	4.14
8	7.57	6.06	5.42	5.05	4.82	4.65	4.53	4.43	4.36	4.30	4.20	4.10	4.00	3.95	3.89	3.84	3.78	3.73	3.67
9	7.21	5.71	5.08	4.72	4.48	4.32	4.20	4.10	4.03	3.96	3.87	3.77	3.67	3.61	3.56	3.51	3.45	3.39	3.33
10	6.94	5.46	4.83	4.47	4.24	4.07	3.95	3.85	3.78	3.72	3.62	3.52	3.42	3.37	3.31	3.26	3.20	3.14	3.08
11	6.72	5.26	4.63	4.28	4.04	3.88	3.76	3.66	3.59	3.53	3.43	3.33	3.23	3.17	3.12	3.06	3.00	2.94	2.88
12	6.55	5.10	4.47	4.12	3.89	3.73	3.61	3.51	3.44	3.37	3.28	3.18	3.07	3.02	2.96	2.91	2.85	2.79	2.72
13	6.41	4.97	4.35	4.00	3.77	3.60	3.48	3.39	3.31	3.25	3.15	3.05	2.95	2.89	2.84	2.78	2.72	2.66	2.60
14	6.30	4.86	4.24	3.89	3.66	3.50	3.38	3.29	3.21	3.15	3.05	2.95	2.84	2.79	2.73	2.67	2.61	2.55	2.49
15	6.20	4.77	4.15	3.80	3.58	3.41	3.29	3.20	3.12	3.06	2.96	2.86	2.76	2.70	2.64	2.59	2.52	2.46	2.40
16	6.12	4.69	4.08	3.73	3.50	3.34	3.22	3.12	3.05	2.99	2.89	2.79	2.68	2.63	2.57	2.51	2.45	2.38	2.32
17	6.04	4.62	4.01	3.66	3.44	3.28	3.16	3.06	2.98	2.92	2.82	2.72	2.62	2.56	2.50	2.44	2.38	2.32	2.25
18	5.98	4.56	3.95	3.61	3.38	3.22	3.10	3.01	2.93	2.87	2.77	2.67	2.56	2.50	2.44	2.38	2.32	2.26	2.19
19	5.92	4.51	3.90	3.56	3.33	3.17	3.05	2.96	2.88	2.82	2.72	2.62	2.51	2.45	2.39	2.33	2.27	2.20	2.13
20	5.87	4.46	3.86	3.51	3.29	3.13	3.01	2.91	2.84	2.77	2.68	2.57	2.46	2.41	2.35	2.29	2.22	2.16	2.09
21	5.83	4.42	3.82	3.48	3.25	3.09	2.97	2.87	2.80	2.73	2.64	2.53	2.42	2.37	2.31	2.25	2.18	2.11	2.04
22	5.79	4.38	3.78	3.44	3.22	3.05	2.93	2.84	2.76	2.70	2.60	2.50	2.39	2.33	2.27	2.21	2.14	2.08	2.00
23	5.75	4.35	3.75	3.41	3.18	3.02	2.90	2.81	2.73	2.67	2.57	2.47	2.36	2.30	2.24	2.18	2.11	2.04	1.97
24	5.72	4.32	3.72	3.38	3.15	2.99	2.87	2.78	2.70	2.64	2.54	2.44	2.33	2.27	2.21	2.15	2.08	2.01	1.94
25	5.69	4.29	3.69	3.35	3.13	2.97	2.85	2.75	2.68	2.61	2.51	2.41	2.30	2.24	2.18	2.12	2.05	1.98	1.91
26	5.66	4.27	3.67	3.33	3.10	2.94	2.82	2.73	2.65	2.59	2.49	2.39	2.28	2.22	2.16	2.09	2.03	1.95	1.88
27	5.63	4.24	3.65	3.31	3.08	2.92	2.80	2.71	2.63	2.57	2.47	2.36	2.25	2.19	2.13	2.07	2.00	1.93	1.85
28	5.61	4.22	3.63	3.29	3.06	2.90	2.78	2.69	2.61	2.55	2.45	2.34	2.23	2.17	2.11	2.05	1.98	1.91	1.83
29	5.59	4.20	3.61	3.27	3.04	2.88	2.76	2.67	2.59	2.53	2.43	2.32	2.21	2.15	2.09	2.03	1.96	1.89	1.81
30	5.57	4.18	3.59	3.25	3.03	2.87	2.75	2.65	2.57	2.51	2.41	2.31	2.20	2.14	2.07	2.01	1.94	1.87	1.79
40	5.42	4.05	3.46	3.13	2.90	2.74	2.62	2.53	2.45	2.39	2.29	2.18	2.07	2.01	1.94	1.88	1.80	1.72	1.64
60	5.29	3.93	3.34	3.01	2.79	2.63	2.51	2.41	2.33	2.27	2.17	2.06	1.94	1.88	1.82	1.74	1.67	1.58	1.48
120	5.15	3.80	3.23	2.89	2.67	2.52	2.39	2.30	2.22	2.16	2.05	1.94	1.82	1.76	1.69	1.61	1.53	1.43	1.31
∞	5.02	3.69	3.12	2.79	2.57	2.41	2.29	2.19	2.11	2.05	1.94	1.83	1.71	1.64	1.57	1.48	1.39	1.27	1.00

Table A4 (Continued)

(c) $F_{\nu_1,\nu_2}(0.01)$

ν_2 \ ν_1	1	2	3	4	5	6	7	8	9	10	12	15	20	24	30	40	60	120	∞
1	4052	4999.5	5403	5625	5764	5859	5928	5981	6022	6056	6106	6157	6209	6235	6261	6287	6313	6339	6366
2	98.50	99.00	99.17	99.25	99.30	99.33	99.36	99.37	99.39	99.40	99.42	99.43	99.45	99.46	99.47	99.47	99.48	99.49	99.50
3	34.12	30.82	29.46	28.71	28.24	27.91	27.67	27.49	27.35	27.23	27.05	26.87	26.69	26.60	26.50	26.41	26.32	26.22	26.13
4	21.20	18.00	16.69	15.98	15.52	15.21	14.98	14.80	14.66	14.55	14.37	14.20	14.02	13.93	13.84	13.75	13.65	13.56	13.46
5	16.26	13.27	12.06	11.39	10.97	10.67	10.46	10.29	10.16	10.05	9.89	9.72	9.55	9.47	9.38	9.29	9.20	9.11	9.02
6	13.75	10.92	9.78	9.15	8.75	8.47	8.26	8.10	7.98	7.87	7.72	7.56	7.40	7.31	7.23	7.14	7.06	6.97	6.88
7	12.25	9.55	8.45	7.85	7.46	7.19	6.99	6.84	6.72	6.62	6.47	6.31	6.16	6.07	5.99	5.91	5.82	5.74	5.65
8	11.26	8.65	7.59	7.01	6.63	6.37	6.18	6.03	5.91	5.81	5.67	5.52	5.36	5.28	5.20	5.12	5.03	4.95	4.86
9	10.56	8.02	6.99	6.42	6.06	5.80	5.61	5.47	5.35	5.26	5.11	4.96	4.81	4.73	4.65	4.57	4.48	4.40	4.31
10	10.04	7.56	6.55	5.99	5.64	5.39	5.20	5.06	4.94	4.85	4.71	4.56	4.41	4.33	4.25	4.17	4.08	4.00	3.91
11	9.65	7.21	6.22	5.67	5.32	5.07	4.89	4.74	4.63	4.54	4.40	4.25	4.10	4.02	3.94	3.86	3.78	3.69	3.60
12	9.33	6.93	5.95	5.41	5.06	4.82	4.64	4.50	4.39	4.30	4.16	4.01	3.86	3.78	3.70	3.62	3.54	3.45	3.36
13	9.07	6.70	5.74	5.21	4.86	4.62	4.44	4.30	4.19	4.10	3.96	3.82	3.66	3.59	3.51	3.43	3.34	3.25	3.17
14	8.86	6.51	5.56	5.04	4.69	4.46	4.28	4.14	4.03	3.94	3.80	3.66	3.51	3.43	3.35	3.27	3.18	3.09	3.00
15	8.68	6.36	5.42	4.89	4.56	4.32	4.14	4.00	3.89	3.80	3.67	3.52	3.37	3.29	3.21	3.13	3.05	2.96	2.87
16	8.53	6.23	5.29	4.77	4.44	4.20	4.03	3.89	3.78	3.69	3.55	3.41	3.26	3.18	3.10	3.02	2.93	2.84	2.75
17	8.40	6.11	5.18	4.67	4.34	4.10	3.93	3.79	3.68	3.59	3.46	3.31	3.16	3.08	3.00	2.92	2.83	2.75	2.65
18	8.29	6.01	5.09	4.58	4.25	4.01	3.84	3.71	3.60	3.51	3.37	3.23	3.08	3.00	2.92	2.84	2.75	2.66	2.57
19	8.18	5.93	5.01	4.50	4.17	3.94	3.77	3.63	3.52	3.43	3.30	3.15	3.00	2.92	2.84	2.76	2.67	2.58	2.49
20	8.10	5.85	4.94	4.43	4.10	3.87	3.70	3.56	3.46	3.37	3.23	3.09	2.94	2.86	2.78	2.69	2.61	2.52	2.42
21	8.02	5.78	4.87	4.37	4.04	3.81	3.64	3.51	3.40	3.31	3.17	3.03	2.88	2.80	2.72	2.64	2.55	2.46	2.36
22	7.95	5.72	4.82	4.31	3.99	3.76	3.59	3.45	3.35	3.26	3.12	2.98	2.83	2.75	2.67	2.58	2.50	2.40	2.31
23	7.88	5.66	4.76	4.26	3.94	3.71	3.54	3.41	3.30	3.21	3.07	2.93	2.78	2.70	2.62	2.54	2.45	2.35	2.26
24	7.82	5.61	4.72	4.22	3.90	3.67	3.50	3.36	3.26	3.17	3.03	2.89	2.74	2.66	2.58	2.49	2.40	2.31	2.21
25	7.77	5.57	4.68	4.18	3.85	3.63	3.46	3.32	3.22	3.13	2.99	2.85	2.70	2.62	2.54	2.45	2.36	2.27	2.17
26	7.72	5.53	4.64	4.14	3.82	3.59	3.42	3.29	3.18	3.09	2.96	2.81	2.66	2.58	2.50	2.42	2.33	2.23	2.13
27	7.68	5.49	4.60	4.11	3.78	3.56	3.39	3.26	3.15	3.06	2.93	2.78	2.63	2.55	2.47	2.38	2.29	2.20	2.10
28	7.64	5.45	4.57	4.07	3.75	3.53	3.36	3.23	3.12	3.03	2.90	2.75	2.60	2.52	2.44	2.35	2.26	2.17	2.06
29	7.60	5.42	4.54	4.04	3.73	3.50	3.33	3.20	3.09	3.00	2.87	2.73	2.57	2.49	2.41	2.33	2.23	2.14	2.03
30	7.56	5.39	4.51	4.02	3.70	3.47	3.30	3.17	3.07	2.98	2.84	2.70	2.55	2.47	2.39	2.30	2.21	2.11	2.01
40	7.31	5.18	4.31	3.83	3.51	3.29	3.12	2.99	2.89	2.80	2.66	2.52	2.37	2.29	2.20	2.11	2.02	1.92	1.80
60	7.08	4.98	4.13	3.65	3.34	3.12	2.95	2.82	2.72	2.63	2.50	2.35	2.20	2.12	2.03	1.94	1.84	1.73	1.60
120	6.85	4.79	3.95	3.48	3.17	2.96	2.79	2.66	2.56	2.47	2.34	2.19	2.03	1.95	1.86	1.76	1.66	1.53	1.38
∞	6.63	4.61	3.78	3.32	3.02	2.80	2.64	2.51	2.41	2.32	2.18	2.04	1.88	1.79	1.70	1.59	1.47	1.32	1.00

Table A4 (Continued)

(d) F_{ν_1, ν_2} (0.005)

ν_2 \ ν_1	1	2	3	4	5	6	7	8	9	10	12	15	20	24	30	40	60	120	∞
1	16211	20000	21615	22500	23056	23437	23715	23925	24091	24224	24426	24630	24836	24940	25044	25148	25253	25359	25465
2	198.5	199.0	199.2	199.2	199.3	199.3	199.4	199.4	199.4	199.4	199.4	199.4	199.4	199.5	199.5	199.5	199.5	199.5	199.5
3	55.55	49.80	47.47	46.19	45.39	44.84	44.43	44.13	43.88	43.69	43.39	43.08	42.78	42.62	42.47	42.31	42.15	41.99	41.83
4	31.33	26.28	24.26	23.15	22.46	21.97	21.62	21.35	21.14	20.97	20.70	20.44	20.17	20.03	19.89	19.75	19.61	19.47	19.32
5	22.78	18.31	16.53	15.56	14.94	14.51	14.20	13.96	13.77	13.62	13.38	13.15	12.90	12.78	12.66	12.53	12.40	12.27	12.14
6	18.63	14.54	12.92	12.03	11.46	11.07	10.79	10.57	10.39	10.25	10.03	9.81	9.59	9.47	9.36	9.24	9.12	9.00	8.88
7	16.24	12.40	10.88	10.05	9.52	9.16	8.89	8.68	8.51	8.38	8.18	7.97	7.75	7.65	7.53	7.42	7.31	7.19	7.08
8	14.69	11.04	9.60	8.81	8.30	7.95	7.69	7.50	7.34	7.21	7.01	6.81	6.61	6.50	6.40	6.29	6.18	6.06	5.95
9	13.61	10.11	8.72	7.96	7.47	7.13	6.88	6.69	6.54	6.42	6.23	6.03	5.83	5.73	5.62	5.52	5.41	5.30	5.19
10	12.83	9.43	8.08	7.34	6.87	6.54	6.30	6.12	5.97	5.85	5.66	5.47	5.27	5.17	5.07	4.97	4.86	4.75	4.64
11	12.23	8.91	7.60	6.88	6.42	6.10	5.86	5.68	5.54	5.42	5.24	5.05	4.86	4.76	4.65	4.55	4.44	4.34	4.23
12	11.75	8.51	7.23	6.52	6.07	5.76	5.52	5.35	5.20	5.09	4.91	4.72	4.53	4.43	4.33	4.23	4.12	4.01	3.90
13	11.37	8.19	6.93	6.23	5.79	5.48	5.25	5.08	4.94	4.82	4.64	4.46	4.27	4.17	4.07	3.97	3.87	3.76	3.65
14	11.06	7.92	6.68	6.00	5.56	5.26	5.03	4.86	4.72	4.60	4.43	4.25	4.06	3.96	3.86	3.76	3.66	3.55	3.44
15	10.80	7.70	6.48	5.80	5.37	5.07	4.85	4.67	4.54	4.42	4.25	4.07	3.88	3.79	3.69	3.58	3.48	3.37	3.26
16	10.58	7.51	6.30	5.64	5.21	4.91	4.69	4.52	4.38	4.27	4.10	3.92	3.73	3.64	3.54	3.44	3.33	3.22	3.11
17	10.38	7.35	6.16	5.50	5.07	4.78	4.56	4.39	4.25	4.14	3.97	3.79	3.61	3.51	3.41	3.31	3.21	3.10	2.98
18	10.22	7.21	6.03	5.37	4.96	4.66	4.44	4.28	4.14	4.03	3.86	3.68	3.50	3.40	3.30	3.20	3.10	2.99	2.87
19	10.07	7.09	5.92	5.27	4.85	4.56	4.34	4.18	4.04	3.93	3.76	3.59	3.40	3.31	3.21	3.11	3.00	2.89	2.78
20	9.94	6.99	5.82	5.17	4.76	4.47	4.26	4.09	3.96	3.85	3.68	3.50	3.32	3.22	3.12	3.02	2.92	2.81	2.69
21	9.83	6.89	5.73	5.09	4.68	4.39	4.18	4.01	3.88	3.77	3.60	3.43	3.24	3.15	3.05	2.95	2.84	2.73	2.61
22	9.73	6.81	5.65	5.02	4.61	4.32	4.11	3.94	3.81	3.70	3.54	3.36	3.18	3.08	2.98	2.88	2.77	2.66	2.55
23	9.63	6.73	5.58	4.95	4.54	4.26	4.05	3.88	3.75	3.64	3.47	3.30	3.12	3.02	2.92	2.82	2.71	2.60	2.48
24	9.55	6.66	5.52	4.89	4.49	4.20	3.99	3.83	3.69	3.59	3.42	3.25	3.06	2.97	2.87	2.77	2.66	2.55	2.43
25	9.48	6.60	5.46	4.84	4.43	4.15	3.94	3.78	3.64	3.54	3.37	3.20	3.01	2.92	2.82	2.72	2.61	2.50	2.38
26	9.41	6.54	5.41	4.79	4.38	4.10	3.89	3.73	3.60	3.49	3.33	3.15	2.97	2.87	2.77	2.67	2.56	2.45	2.33
27	9.34	6.49	5.36	4.74	4.34	4.06	3.85	3.69	3.56	3.45	3.28	3.11	2.93	2.83	2.73	2.63	2.52	2.41	2.29
28	9.28	6.44	5.32	4.70	4.30	4.02	3.81	3.65	3.52	3.41	3.25	3.07	2.89	2.79	2.69	2.59	2.48	2.37	2.25
29	9.23	6.40	5.28	4.66	4.26	3.98	3.77	3.61	3.48	3.38	3.21	3.04	2.86	2.76	2.66	2.56	2.45	2.33	2.21
30	9.18	6.35	5.24	4.62	4.23	3.95	3.74	3.58	3.45	3.34	3.18	3.01	2.82	2.73	2.63	2.52	2.42	2.30	2.18
40	8.83	6.07	4.98	4.37	3.99	3.71	3.51	3.35	3.22	3.12	2.95	2.78	2.60	2.50	2.40	2.30	2.18	2.06	1.93
60	8.49	5.79	4.73	4.14	3.76	3.49	3.29	3.13	3.01	2.90	2.74	2.57	2.39	2.29	2.19	2.08	1.96	1.83	1.69
120	8.18	5.54	4.50	3.92	3.55	3.28	3.09	2.93	2.81	2.71	2.54	2.37	2.19	2.09	1.98	1.87	1.75	1.61	1.43
∞	7.88	5.30	4.28	3.72	3.35	3.09	2.90	2.74	2.62	2.52	2.36	2.19	2.00	1.90	1.79	1.67	1.53	1.36	1.00

Appendix:

Table A4 (Continued)

(e) $F_{\nu_1,\nu_2}(0.001)$

ν_2 \\ ν_1	1	2	3	4	5	6	7	8	9	10	12	15	20	24	30	40	60	120	∞
1	4053*	5000*	5404*	5625*	5764*	5859*	5929*	5981*	6023*	6056*	6107*	6158*	6209*	6235*	6261*	6287*	6313*	6340*	6366*
2	998.5	999.0	999.2	999.2	999.3	999.3	999.4	999.4	999.4	999.4	999.4	999.4	999.4	999.5	999.5	999.5	999.5	999.5	999.5
3	167.0	148.5	141.1	137.1	134.6	132.8	131.6	130.6	129.9	129.2	128.3	127.4	126.4	125.9	125.4	125.0	124.5	124.0	123.5
4	74.14	61.25	56.18	53.44	51.71	50.53	49.66	49.00	48.47	48.05	47.41	46.76	46.10	45.77	45.43	45.09	44.75	44.40	44.05
5	47.18	37.12	33.20	31.09	29.75	28.84	28.16	27.64	27.24	26.92	26.42	25.91	25.39	25.14	24.87	24.60	24.33	24.06	23.79
6	35.51	27.00	23.70	21.92	20.81	20.03	19.46	19.03	18.69	18.41	17.99	17.56	17.12	16.89	16.67	16.44	16.21	15.99	15.75
7	29.25	21.69	18.77	17.19	16.21	15.52	15.02	14.63	14.33	14.08	13.71	13.32	12.93	12.73	12.53	12.33	12.12	11.91	11.70
8	25.42	18.49	15.83	14.39	13.49	12.86	12.40	12.04	11.77	11.54	11.19	10.84	10.48	10.30	10.11	9.92	9.73	9.53	9.33
9	22.86	16.39	13.90	12.56	11.71	11.13	10.70	10.37	10.11	9.89	9.57	9.24	8.90	8.72	8.55	8.37	8.19	8.00	7.81
10	21.04	14.91	12.55	11.28	10.48	9.92	9.52	9.20	8.96	8.75	8.45	8.13	7.80	7.64	7.47	7.30	7.12	6.94	6.76
11	19.69	13.81	11.56	10.35	9.58	9.05	8.66	8.35	8.12	7.92	7.63	7.32	7.01	6.85	6.68	6.52	6.35	6.17	6.00
12	18.64	12.97	10.80	9.63	8.89	8.38	8.00	7.71	7.48	7.29	7.00	6.71	6.40	6.25	6.09	5.93	5.76	5.59	5.42
13	17.81	12.31	10.21	9.07	8.35	7.86	7.49	7.21	6.98	6.80	6.52	6.23	5.93	5.78	5.63	5.47	5.30	5.14	4.97
14	17.14	11.78	9.73	8.62	7.92	7.43	7.08	6.80	6.58	6.40	6.13	5.85	5.56	5.41	5.25	5.10	4.94	4.77	4.60
15	16.59	11.34	9.34	8.25	7.57	7.09	6.74	6.47	6.26	6.08	5.81	5.54	5.25	5.10	4.95	4.80	4.64	4.47	4.31
16	16.12	10.97	9.00	7.94	7.27	6.81	6.46	6.19	5.98	5.81	5.55	5.27	4.99	4.85	4.70	4.54	4.39	4.23	4.06
17	15.72	10.66	8.73	7.68	7.02	6.56	6.22	5.96	5.75	5.58	5.32	5.05	4.78	4.63	4.48	4.33	4.18	4.02	3.85
18	15.38	10.39	8.49	7.46	6.81	6.35	6.02	5.76	5.56	5.39	5.13	4.87	4.59	4.45	4.30	4.15	4.00	3.84	3.67
19	15.08	10.16	8.28	7.26	6.62	6.18	5.85	5.59	5.39	5.22	4.97	4.70	4.43	4.29	4.14	3.99	3.84	3.68	3.51
20	14.82	9.95	8.10	7.10	6.46	6.02	5.69	5.44	5.24	5.08	4.82	4.56	4.29	4.15	4.00	3.86	3.70	3.54	3.38
21	14.59	9.77	7.94	6.95	6.32	5.88	5.56	5.31	5.11	4.95	4.70	4.44	4.17	4.03	3.88	3.74	3.58	3.42	3.26
22	14.38	9.61	7.80	6.81	6.19	5.76	5.44	5.19	4.99	4.83	4.58	4.33	4.06	3.92	3.78	3.63	3.48	3.32	3.15
23	14.19	9.47	7.67	6.69	6.08	5.65	5.33	5.09	4.89	4.73	4.48	4.23	3.96	3.82	3.68	3.53	3.38	3.22	3.05
24	14.03	9.34	7.55	6.59	5.98	5.55	5.23	4.99	4.80	4.64	4.39	4.14	3.87	3.74	3.59	3.45	3.29	3.14	2.97
25	13.88	9.22	7.45	6.49	5.88	5.46	5.15	4.91	4.71	4.56	4.31	4.06	3.79	3.66	3.52	3.37	3.22	3.06	2.89
26	13.74	9.12	7.36	6.41	5.80	5.38	5.07	4.83	4.64	4.48	4.24	3.99	3.72	3.59	3.44	3.30	3.15	2.99	2.82
27	13.61	9.02	7.27	6.33	5.73	5.31	5.00	4.76	4.57	4.41	4.17	3.92	3.66	3.52	3.38	3.23	3.08	2.92	2.75
28	13.50	8.93	7.19	6.25	5.66	5.24	4.93	4.69	4.50	4.35	4.11	3.86	3.60	3.46	3.32	3.18	3.02	2.86	2.69
29	13.39	8.85	7.12	6.19	5.59	5.18	4.87	4.64	4.45	4.29	4.05	3.80	3.54	3.41	3.27	3.12	2.97	2.81	2.64
30	13.29	8.77	7.05	6.12	5.53	5.12	4.82	4.58	4.39	4.24	4.00	3.75	3.49	3.36	3.22	3.07	2.92	2.76	2.59
40	12.61	8.25	6.60	5.70	5.13	4.73	4.44	4.21	4.02	3.87	3.64	3.40	3.15	3.01	2.87	2.73	2.57	2.41	2.23
60	11.97	7.76	6.17	5.31	4.76	4.37	4.09	3.87	3.69	3.54	3.31	3.08	2.83	2.69	2.55	2.41	2.25	2.08	1.89
120	11.38	7.32	5.79	4.95	4.42	4.04	3.77	3.55	3.38	3.24	3.02	2.78	2.53	2.40	2.26	2.11	1.95	1.76	1.54
∞	10.83	6.91	5.42	4.62	4.10	3.74	3.47	3.27	3.10	2.96	2.74	2.51	2.27	2.13	1.99	1.84	1.66	1.45	1.00

* Multiply these entries by 100

Answers to Exercises

CHAPTER 2

Section A

1. Mean $= 0.4283$ g; variance $= 8.3668 \times 10^{-4}$; s.d. $= 0.02893$.

2. Mean $= 263.75$; variance $= 141.071$; s.d. $= 11.877$.

3. Mean $= 0.3313$; variance $= 1.05132 \times 10^{-5}$; s.d. $= 3.2424 \times 10^{-3}$.

4.

x	335	340	345	350	355	360	365
f	1	3	5	9	8	3	1

Mean $= 350.5$; variance $= 47.1552$; s.d. $= 6.8670$.

5. Mean $= 9.2931$; variance $= 0.024780$; s.d. $= 0.15742$.

6.

x	8.7	8.8	8.9	9.0	9.1	9.2	9.3	9.4	9.5	9.6	9.7	9.8	9.9	10.0	10.1	10.2	10.3
f	1	1	0	1	0	3	1	1	1	4	3	6	7	5	9	6	5

Mean $= 9.8426$; variance $= 0.14740$; s.d. $= 0.3839$.

7.

x	0.3295	0.3495	0.3695	0.3895	0.4095	0.4295	0.4495	0.4695	0.4895
f	1	0	8	10	22	23	29	6	1

Mean $= 0.4241$; variance $= 8.6347 \times 10^{-4}$; s.d. $= 0.02938$.

8. (i) 1st quartile $= 0.412$; median $= 0.436$; 3rd quartile $= 0.445$.
   (ii) 1st quartile $= 253$; median $= 265$; 3rd quartile $= 274$.
   (iii) 1st quartile $= 0.328$; median $= 0.332$; 3rd quartile $= 0.334$.
   (Same answer (to 3D) is obtained in (iii) whether we treat the observations as a sample or frequency distribution.)

9. 10th percentile $= 340.8$; median $= 350.8$; 90th percentile $= 359.2$.

10. (i) 9.00, 9.19, 9.30, 9.40, 9.54; SIQR = 0.105.
    (ii) 0.3695, 0.4050, 0.4273, 0.4471, 0.4662; SIQR = 0.021.

11. (a) Quartiles for example 6   9.6̇, 9.92, 10.12
         Quartiles for example 7   0.4050, 0.4273, 0.4471

    (b) example 6

         | 8. | 7 |
         | 8. | 8 |
         | 9. | 0 |
         | 9. | 2 2 2 |
         | 9. | 3 |
         | 9. | 4 |
         | 9. | 5 |
         | 9. | 6 6 6 6 |
         | 9. | 7 7 7 |
         | 9. | 8 8 8 8 8   8 |
         | 9. | 9 9 9 9 9   9 9 |
         | 10. | 0 0 0 0 0 |
         | 10. | 1 1 1 1 1   1 1 1 1 |
         | 10. | 2 2 2 2 2   2 |
         | 10. | 3 3 3 3 3 |

    example 7

0.32	7					
0.36	2 2 6 6 7	10 16 19				
0.38	3 8 8 10 10	11 11 11 12 14				
0.40	0 2 5 7 8	10 10 12 12 13	13 14 14 14 15	16 16 17 19 19	19 19	
0.42	2 2 3 4 4	5 5 5 6 6	7 7 8 9 10	13 14 15 16 18	18 19 19	
0.44	0 0 0 1 1	2 2 3 3 3	4 4 5 5 5	5 6 6 6 8	8 10 10 12 13	13 14 18 18
0.46	1 4 4 6 9	18				
0.48	5					

12. Mean = 0.4276 g; variance = 9.3253×10^{-4}.

Section B

1. (i) Angles   0.65, 0.1780; 0.552, 0.1810; 0.668, 0.1965; 0.638, 0.1998;
               0.548, 0.1681; 0.420, 0.1349; 0.186, 0.0261.
   (ii) Metals   0.7329, 0.2545; 0.6257, 0.2281; 0.4186, 0.1141; 0.3971,
               0.1331; 0.4414, 0.1525.
   (iii) 0.523, 0.2197.

2. 32.892, 0.37.

3. | x | 14.5 | 24.5 | 34.5 | 44.5 | 54.5 | 64.5 | 74.5 | 84.5 | 94.5 | 104.5 | 114.5 |
   |-----|------|------|------|------|------|------|------|------|------|-------|-------|
   | f | 2 | 3 | 3 | 6 | 9 | 8 | 7 | 8 | 1 | 1 | 2 |

   $\bar{x} = 62.1, s = 23.44$.

4.

x	6.545	6.645	6.745	6.845	6.945	7.045	7.145	7.245	7.345	7.445
f	2	5	6	14	26	18	13	10	3	3

$\bar{x} = 6.997, s = 0.1936$.

5. 0.80, 0.63, 0.94; 0.73, 0.48, 0.75; 0.46, 0.35, 0.50; 0.45, 0.33, 0.51; 0.45, 0.31, 0.58.

6. 6.983, 6.745, 7.255.

7. (a) Quartiles example 3   47, 62, 78.7857.
   Quartiles example 4   6.881, 6.983, 7.126.

   (b) example 3
   ```
   10      7 9
   20      5 6 7
   30      0 8 9
   40      1 2 3 6 7  9
   50      0 3 3 4 4  5 6 8 9
   60      0 2 3 3 4  5 6 8
   70      1 3 4 4 7  8 9
   80      0 2 2 5 5  7 7 9
   90      8
  100      9
  110      0 8
   ```

   example 4
   ```
   6.5  4 6
   6.6  0 2 3 4 9
   6.7  1 3 6 6 7  9
   6.8  0 0 1 2 2  4 5 5 5 7  8 8 9 9
   6.9  0 0 1 1 1  2 2 2 2 3  4 4 4 5 5  6 6 6 6 7  7 8 8 9 9   9
   7.0  0 1 1 1 2  2 3 3 4 5  5 5 6 7 7  8 8 9
   7.1  0 0 2 2 4  4 4 5 6 6  7 7 9
   7.2  1 2 4 4 4  5 6 6 8 8
   7.3  3 6 8
   7.4  0 2 5
   ```

8. 77.2$\dot{6}$, 10.2013.

CHAPTER 3

Section A

1. (a) 0.63   (b) 0.06   (c) 0.74   (d) 0   (e) 0.69   (f) 0.45   (g) 0.92.

2. (a) 8/15   (b) 0   (c) 7/15.

3. (a) 0.15   (b) 0.35   (c) 0.65.

4. (a) 0.3   (b) 0.5   (c) 0.1   (d) 0.2 (e) 0.7.

5. (a) 0.72   (b) 0.26   (c) 0.02.

6. 8/33, 16/33, 19/33. 28/165, 28/55, 41/55.

7. 0.2076, B.

Section B
1. (a) 7/30   (b) 14/30   (c) 2/30.

2. 0.027

3. 209/343.

4. 0.9702. 0.3356

5. 0.0199. 109/199.

6. 0.2.

CHAPTER 4

Section A
1. 0.1382; 4.

2. (a) 0.0060 (b) 0.9536 (c) 0.0123.

3. 0.03125 (i) 0.7280 (ii) 8.9×10^{-16}.

4. 14.2, 28.5, 23.7, 10.5, 2.6, 0.4, 0.0.

5. (a) 0.6065 (b) 0.0902 (c) 0.9982.

6. 0.3233; 1 or 2.

7. $\bar{x} = 1.435, s^2 = 1.3927; 47.6, 68.3, 49.0, 23.5, 8.4, 2.4, 0.6$.

8. $\bar{x} = 1.5, s^2 = 1.5238; 89.3, 133.9, 100.4, 50.2, 18.8, 5.6, 1.4$.

9. (i) 0.7568 (ii) 0.0205 (iii) 0.0456 (iv) 0.5948.

10. (i) 0 (ii) 1.64 (iii) 3.29 (iv) −3.09.

11. (i) 0.9838 (ii) 0.9836 (iii) 0.0021 (iv) 0 if 350 is exact *or* 0.0566 if $X = 350$ means $349.5 < X < 350.5$.
    $Z = (375 - 345)/7 = 4.29$. $\Pr(Z > 4.29)$ is very small so that mean is not likely to be 345.

12. 283, 8.

13. $\mu = 350; \sigma = 10$.

14. (a) 0.9073 (b) 0.0905 (c) 0.0023.

15. 349.75, 6.0459. 4.5, 14.8, 29.2, 30.1, 16.1, 4.6, 0.7.

16. 540.5, 10, 14%.

17. 188 (use normal approximation to the binomial).

18. (a) 0.3874 (b) 0.9666 (c) 0.9666.

19. 0.0228, 0.6645.

20. 0.128, 0.64.

21. (a) 0.7769 (b) 0.1733. 0.0112.

Section B

1. 1/256, 12/256, 54/256, 108/256, 81/256; 3, 3/4.

2. (a) 5.905×10^{-16} (b) 0.0317 (c) 0.9972.

3. (a) 0.3487 (b) 0.0702 (c) 9.

4. 3.9997×10^{-12}.

5. (a) 0.7738 (b) 3.125×10^{-7} (c) 0.0226.

6. 773.8, 203.6, 21.4, 1.1, 0.0, 0.0; 5%.

7. 0.04979, 0.14936, 0.22404, 0.22404, 0.16803, 0.10082, 0.05041, 0.02610, 0.00810, 0.00270, 0.00081, 0.00022, 0.00006, 0.00001.

8. 0.8571; 1 or 2.

9. 0.607, 33.

10. 4 or 5; 0.00709, 0.01892.

11. 73.6, 73.6, 36.8, 12.3, 3.1, 0.6.

12. (i) 0.7045 (ii) 0.1644 (iii) 0.6826 (iv) 0.8925.

13. (i) 0 (ii) -1.64 (iii) 2.58 (iv) 3.09.

14. (i) 0.9236 (ii) 0.9958 (iii) 0.0162.
    (iv) 0 if 350 is exact; 0.0566 if $X = 350$ means $349.5 < X < 350.5$.

15. (a) 7.42 (b) 7.45 (c) 7.48.

16. $\mu = 94.5, \sigma = 4.6$.

17. (a) 0.9073 (b) 0.0905 (c) 0.0023.

18. 0.1234 (binomial); 0.1306 (normal approx. to binomial).

19. (a) 0.0571 (b) 0.9198 (c) 0.6424 (d) 0.0078.

20. $\bar{x} = 288.5084, s = 15.8767$.
    6.2, 6.6, 12, 18.4, 27.4, 34.6, 42.2, 43.9, 44, 37.5, 30.9, 21.7, 14.6, 8.4,
    4.7, 2.2, 1.6.

21. (a) 0.09 (b) 0.2195; 7.

22. (a) 0.0672 (b) 0.1834 (c) 0.0496.

CHAPTER 5

Section A

1. $\bar{X} \sim N\left(9.30; \dfrac{0.15^2}{36}\right)$ (a) 0.4772 (b) 0.0548 (c) 0.6554 (d) nearly 1.

2. (a) 0.3303, 0.3323 (b) 0.3296, 0.3330.

3. $z = 2.76$; significant at the 1% level (two-tailed test).

4. $z = 1.33$; not significant (one-tailed test).

5. 0.3314; 0.3305, 0.3324

6. $z = 9.57$; significant at the 0.1% level (one-tailed test); 16.5, 24.9.

7. $z = 1.5$; not significant (one-tailed test); 0.091, 0.209.

8. 5.52, 38.85; $F = 1.86, v_1 = 9, v_2 = 11$. N.S.; $t = 0.46, v = 9$, N.S.
   $t = 4.175, v = 11$; significant at the 0.1% level; $t = 1.99, v = 20$, N.S.
   (all two-tailed tests).

9. $t = -3.32$ on 9 d.f.s; significant at the 1% level (one-tailed test).
   12.3, 15.7.

10. (a) $t = 11.29$ on 19 d.f.; significant at the 0.1% level (one-tailed test).
    (b) $z = 12.34$; significant at the 0.1% level (one-tailed test).
    (c) $t = 7.76$ on 33 d.f.; significant at the 0.1% level (one-tailed test).

11. $t = 1.71$ on 7 d.f.; not significant (two-tailed test).

12. (a) $\chi^2 = 5.52, v = 2; \chi_2^2 (0.05) = 5.99$. Fit is just satisfactory.
    (b) $\chi^2 = 0.22, v = 3; \chi_3^2 (0.05) = 7.81$. Fit is satisfactory.
    (c) $\chi^2 = 2.39, v = 2; \chi_2^2 (0.05) = 5.99$. Fit is satisfactory.

13. 333.80, 336.60; $\Pr(\mu > 330) = \Pr(Z > -7.262) = 1.0000$.

Section B

1. (a) 0.9642 (b) 0.0059 (c) 0.0013.

2. 11.82, 12.18.

3. $z = -2$; significant at the 5% level (one-tailed test).

4. 0.3314; 0.3305, 0.3324.

5. $z = 15.36$; significant at 0.1% level (two-tailed test).
   (a) $z = 4.85$; significant at 0.1% level (two-tailed test).
   (b) $z = 17.03$; significant at 0.1% level (two-tailed test).

6. $z = -1.30$; N.S. (one-tailed test).

7. $z = 1.27$; N.S. (two-tailed test).

8. $\chi^2 = 38.66, \nu = 29; \chi^2_{29}(0.05) = 42.56$; N.S.

9. $F = 1.062, \nu_1 = 59, \nu_2 = 63$; N.S.

10. (a) 0.0087 (b) 0.8882. $F = 1.23, \nu_1 = 24, \nu_2 = \infty$; N.S. $t = -1.43, \nu = 24$ N.S. (two-tailed test).

11. $t = 0.57, \nu = 19$, N.S. (two-tailed test); $t = 0.40, \nu = 38$, N.S. (two-tailed test).

12. $-0.20, 0.73; t = 1.28, \nu = 10$, N.S. (one-tailed test).

13. $t = 1.48, \nu = 4$, N.S. (two-tailed test).

14. (a) $\chi^2 = 0.45, \nu = 1; \chi^2_1(0.05) = 3.84$. Fit is satisfactory.
   (b) $\chi^2 = 1.39, \nu = 2; \chi^2_2(0.05) = 5.99$. Fit is satisfactory.
   (c) $\chi^2 = 35.34, \nu = 12; \chi^2_{12}(0.05) = 21.03$. Fit is poor.

15. 468.1, 511.9. $\Pr(\mu < 500) = \Pr(Z < 0.89) = 0.81$.

CHAPTER 6

Section A

1. $y = -2.65 + 0.0337x$; $F = 732.84, \nu_1 = 1, \nu_2 = 4$, significant at the 0.1% level.
   (a) 0.21 (b) 0.45.

2. $y = -0.2281 + 0.9948x$. 0.9827, 1.0068.

3. $y = 8.65 + 0.30x$. 11.35; 11.104, 11.596.

4. $y = 56.83 + 1.5989x$. $r = 0.9391$; $t = 6.692, \nu = 6$, significant at the 0.1% level (one-tailed test).

5. $y = -1.488 + 1.191x; x = 1.427 + 0.6197y$. $r = 0.8593, t = 3.36, v = 4$.
    Significant at the 5% level (two-tailed test).

6. $\ln y = 3.6889 + 0.3699x; \hat{y}_8 = 771$.

x	0	1	2	3	4	5	6	7
$\hat{y}$	3.69	4.06	4.43	4.80	5.17	5.54	5.91	6.28
lower limit	3.30	3.74	4.17	4.58	4.95	5.28	5.59	5.89
upper limit	4.08	4.38	4.68	5.02	5.39	5.79	6.22	6.67

Section B

1. $y = 3.214x - 1.829$.

2. $D = 1.35 + 0.234v^2$.

3. $y = 61.28 + 7.714x$.   6.64, 8.79.   130.7.

4. $y = 20 + 4.82x$.   3.86, 5.78.

5. $S = 0.5140 + 0.0028T$.   0.0021, 0.0035.

6. $y = 49.7x + 87.8$; $x = 0.0131y - 0.32$. $r = 0.8075$, $t = 3.35$, $v = 6$.
    Significant at the 1% level (one-tailed test).

7. $y = 1.7302 + 0.2611t$. (where $y = (\text{conductivity})^2 / 10^6$)   0.2215, 0.3007.

8. (b) $y = 17.8552 + 11.6171x$.
    (c) 35.8213 (for least squares line).
    (d) $F = 144.81$, $v_1 = 1$, $v_2 = 3$. Significant at the 1% level.
    (e) $t = 12.03$, $v = 3$. Significant at the 1% level.
    (f) (i) 5.98, 17.26 (ii) 0.52, 35.19 (iii) 16.61, 42.33.

Index